LANDFORMS AND GEOLOGY OF GRANITE TERRAINS

LANDFORMS AND GEOLOGY OF GRANITE TERRAINS

C.R. Twidale
University of Adelaide, Australia

J.R. Vidal Romaní
University of Coruña, Spain

CRC Press is an imprint of the
Taylor & Francis Group, an **informa** business

A BALKEMA BOOK

Library of Congress Cataloging-in-Publication Data
Applied for

Cover Illustrations
Front: Acuminate and ensiform residual, Mt. Manypeaks area, near Albany, Australia.
Photograph by C.R. Twidale.
Back: Domed granite inselberg from the Bolson of the Señor de la Peña Valley, Anillaco,
República Argentina. Photograph by J.R. Vidal Romaní.

Copyright © 2005 Taylor & Francis Group plc, London, UK

All rights reserved. No part of this publication or the information contained herein may be reproduced, stored in a retrieval system, or transmitted in any form or by any means, electronic, mechanical, by photocopying, recording or otherwise, without written prior permission from the publisher.

Although all care is taken to ensure the integrity and quality of this publication and the information herein, no responsibility is assumed by the publishers nor the author for any damage to property or persons as a result of operation or use of this publication and/or the information contained herein.

Published by: A.A. Balkema Publishers Leiden, The Netherlands, a member of
Taylor & Francis Group plc
www.balkema.nl and www.tandf.co.uk

ISBN 04 1536 435 3

Table of Contents

Preface and acknowledgements		IX
1	**Characteristics and foundations**	**1**
	1.1 Typical landforms and landscapes	1
	1.2 Previous work	5
	1.3 Occurrences of granite	9
	1.4 Granite – definition and composition	10
	1.5 Physical characteristics	13
	1.6 Orthogonal fracture systems	16
	1.7 Fractures and drainage patterns	18
2	**Sheet fractures and structures**	**27**
	2.1 Terminology	27
	2.2 Description and characteristics	28
	2.3 Theories of origin	32
	2.3.1 Exogenetic explanations	33
	2.3.2 Endogenetic explanations	41
	2.4 Summary	47
3	**Weathering**	**49**
	3.1 Definition and significance	49
	3.2 Physical disintegration	49
	3.3 Chemical alteration	54
	3.4 The course of weathering in granite	54
	3.5 Controls of weathering	58
4	**Plains – the expected granite form**	**63**
	4.1 Weathering and surfaces of low relief	63
	4.2 Plains of epigene (subaerial) origin	63
	4.2.1 Rolling or undulating plains	64
	4.2.2 Pediments	66
	4.2.3 Relationship between pediment and peneplain	72
	4.3 Etch plains in granite	72
	4.4 Very flat plains	74
	4.5 Multicyclic and stepped assemblages	75
	4.6 Exhumed plains	78
	4.7 Summary	79
5	**Boulders as examples of two-stage forms**	**81**
	5.1 The two-stage or etching mechanism	81
	5.2 Boulders – morphology and occurrences	83
	5.3 Subsurface exploitation of orthogonal fracture systems	83
	5.4 Tectonic and structural forms	95
	5.5 Types of peripheral or marginal weathering	97

	5.6	Causes of peripheral weathering	98
	5.7	Evacuation of grus	102
	5.8	Boulders of epigene origin	103
	5.9	Summary	105

6 Inselbergs and bornhardts — 109
   6.1 Definitions and terminology   109
   6.2 Bornhardt characteristics   115
   6.3 Theories of origin   118
       6.3.1 Environment   118
       6.3.2 The scarp retreat hypothesis   120
       6.3.3 Tectonics and structure: faulting and lithology   121
       6.3.4 Variations in fracture density   127
       6.3.5 Differential subsurface weathering and the two-stage concept   127
   6.4 Evidence and argument concerning origins of bornhardts   131
       6.4.1 Contrasts in weathering between hill and plain   131
       6.4.2 Incipient domes   131
       6.4.3 Subsurface initiation of minor forms   131
       6.4.4 Flared slopes and stepped inselbergs   132
       6.4.5 Regional and local patterns in plan   139
       6.4.6 Coexistence of forms associated with compression/shearing   140
       6.4.7 Topographic settings   140
       6.4.8 Occurrence in multicyclic landscapes   140
       6.4.9 Fracture-defined margins   142
       6.4.10 Age of inselbergs and bornhardts   142
   6.5 Exhumed bornhardts and inselbergs   146
   6.6 Antiquity and inselberg landscapes   149
   6.7 Summary   149

7 Other granitic residuals and uplands — 153
   7.1 Isolated residuals   153
       7.1.1 Nubbins   153
       7.1.2 Castle koppies   155
       7.1.3 Large conical forms or medas   159
       7.1.4 Towers and acuminate forms   162
   7.2 Massifs   163
   7.3 Regions of all slopes topography   169
   7.4 Discussion   171

8 Minor forms developed on steep slopes — 173
   8.1 Flared slopes   173
       8.1.1 Description and characteristics   173
       8.1.2 Origin   177
       8.1.3 Changes after exposure   183
   8.2 Fretted basal slopes and other variants   184
   8.3 Scarp-foot weathering and erosion, and the piedmont angle   188
   8.4 Rock platforms   190
       8.4.1 Description   190
       8.4.2 Origin   190

	8.5	Scarp-foot depressions	190
		8.5.1 Description	190
		8.5.2 Origin	192
	8.6	Flutings or grooves	193
		8.6.1 Description	193
		8.6.2 Origin	196
		8.6.3 Surface or subsurface initiation?	200
		8.6.4 Inversion	203

9 Minor forms developed on gentle slopes — 207

	9.1	Rock basins	207
		9.1.1 Description	207
		9.1.2 Nomenclature	211
		9.1.3 Origin	211
		9.1.4 Differentiation of major types	216
		9.1.5 Evacuation of debris	219
		9.1.6 Rate of development	219
	9.2	Plinths and associated blocks and boulders	220
		9.2.1 Description	220
		9.2.2 Origin	221
	9.3	Pedestal rocks	222
		9.3.1 Terminology	222
		9.3.2 Origin	222
	9.4	Gutters or runnels	224
		9.4.1 Terminology	224
		9.4.2 Description	224
		9.4.3 Origin	226
	9.5	Rock levees	228
	9.6	Rock doughnuts	230
		9.6.1 Description	230
		9.6.2 Origin	231
		9.6.3 Evidence and argument	232
	9.7	Fonts	232

10 Caves and tafoni — 235

	10.1	General statement	235
	10.2	Caves associated with corestones and grus	235
	10.3	Caves associated with fractures	236
	10.4	Tafoni	238
		10.4.1 Description	238
		10.4.2 Process	245
		10.4.3 Stages of development	249
		10.4.4 Case-hardening and other veneers	250
	10.5	Speleothems	251

11 Split and cracked blocks and slabs — 259

	11.1	Split rocks	259
		11.1.1 Description	259
		11.1.2 Origin	260
	11.2	Parted and dislodged blocks	264
	11.3	Dislocated slabs	266
		11.3.1 A-tents	266

VIII Table of Contents

		11.3.2	Overlapping slabs	271
		11.3.3	Displaced slabs	271
		11.3.4	Chaos	275
		11.3.5	Wedges	276
		11.3.6	Origin of the forms	278
		11.3.7	Relationship of A-tents and pressure ridges	282
	11.4	Polygonal cracking		282
		11.4.1	Description	282
		11.4.2	Previous interpretations	285
		11.4.3	Evidence	287
		11.4.4	Explanations	287
	11.5	Tesselated pavements		291
12	**Zonality, azonality and the coastal context**			**293**
	12.1	Introduction		293
	12.2	Lithological zonality and azonality		293
	12.3	Climatic zonality and azonality		309
	12.4	The coastal context		313
13	**Retrospect and prospect**			**327**
	Author index			331
	Location index			335
	Subject index			343
	About the authors			352

Preface and acknowledgements

The authors owe a great debt of gratitude to many individuals and organisations without whose support and assistance the research, and particularly the field investigations, on which this book is based could not have been carried out. The Australian Research Council (and its several predecessors) have supported investigations of various aspects of granite landform evolution. The universities of Adelaide and of Coruña have also provided support and facilities, and through study leave and similar schemes, supported the travel, consultation and collaboration without which this book would not have been possible.

It is impossible to acknowledge specifically the interest and encouragement of the many individuals, in various parts of the world, who have given us part of their time, as well as the benefit of their local knowledge; but we thank them all. Special thanks are due to Dr Liz Campbell and Dr Jennie Bourne, both of whom have, over a period of many years, been enthusiastic, lively and critical field collaborators and companions, as well as co-authors of many of the papers that form the foundation of this larger work. Mr Peter Moss, Dr Liz Campbell, Mrs Noreen Shepherd and Ana Martelli that read the book in draft form and helped eliminate many errors and inconsistencies. Debbie Haggar is responsible for drafting many of the line drawings, Jacie Davis for reproducing them.

1

Characteristics and foundations

1.1 TYPICAL LANDFORMS AND LANDSCAPES

Many familiar landforms are developed on granite as well as in other lithological environments. Thus, fault scarps and fault-line scarps, valleys (Figs 1.1a, b, and c) and other features (Figs 1.1d and e) are

(a)

Figure 1.1. (a) The Meckering fault scarp, Yilgarn Province, Western Australia was a complex of dirt scarps in granitic rocks, and formed on October 20, 1968. All the scarps are now degraded (West Australian Newspapers).

2 *Landforms and Geology of Granite Terrains*

(b)

(c)

Figure 1.1. (b) The MacDonald fault-line scarp in northwestern Canada, with granite underlying the higher ground, sediments the lower (Department of Energy Mines and Resources, Canada). (c) Fault-line valley in granitic terrain, Kazan region, northwestern Canada. The fracture has been intruded by a quartz vein (Department of Energy, Mines and Resources, Canada).

Characteristics and foundations 3

Figure 1.1. (d) Granite platform, central Namibia, with traces of orthogonal fracture system, but with the edges of the blocks defined by raised rims, due to toughening by recrystallisation associated with fault dislocation. (e) Fracture system with raised rims due to crystallisation of infilling fluids (quartz) disintegration of the rock mass. Cavallers Dam, Pyrenees, northeastern Spain.

as well-developed in granite as elsewhere, glaciated granite landscapes bear the clear and characteristic imprint of ice sheets and glaciers (Fig. 1.1f), and where capped by a resistant carapace of laterite or silcrete, for example, plateau forms are developed (Fig. 1.1g). No landform is entirely peculiar to, and developed only on, granite. All are also found on outcrops of other rocks (see Chapter 13).

On the other hand, numerous landforms, major and minor, are more commonly developed on granite than on any other rock type, and can thus be regarded as characteristic of such terrains. Indeed, many of the landforms and landform assemblages developed on granite are so distinctive as to allow a high probability identification of granite exposures on the basis of distant views, or

4 Landforms and Geology of Granite Terrains

Figure 1.1. (f) Glaciated landscape in the Sierra Nevada of California (United States Geological Survey). (g) Duricrusted mesa in northwest Queensland with laterite developed on granite (CSIRO) Australia.

of inspection of air photographs. Boulders and bornhardts, koppies and nubbins, and a range of minor forms, notably such features as basins and flutings, are characteristic of granitic terrains (Figs 1.3a, b, c, d, e and f). Granite hills, and particularly bornhardts, standing in isolation as inselbergs, together with the plains that are the most extensive granite form, give rise to the inselberg landscapes or Inselberglandschaften that so captured the imaginations of early travellers (Toit, 1954) in eastern and southern Africa, India and central Australia (Figs 1.2a and b).

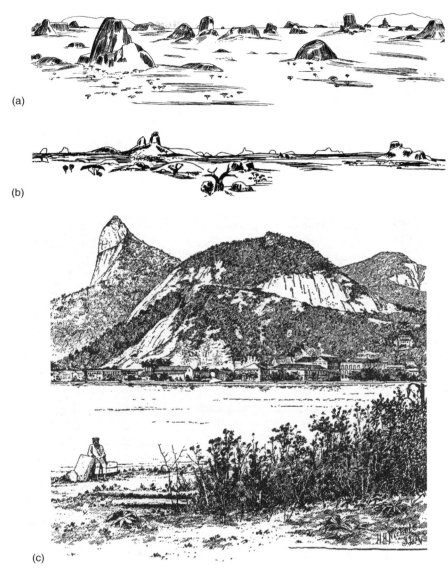

Figure 1.2. (a) and (b) Jessen's sketches of inselberg landscapes in Angola (Jessen, 1936). (c) Sketch of morros near Rio de Janeiro, Brazil, (Branner, 1896).

1.2 PREVIOUS WORK

Granitic landforms and terrains have excited the interest of geologists, geographers and geomorphologists for more than two centuries, resulting in a voluminous journal literature and brilliant sketches (Fig. 1.2). Boulders and pediments, bornhardts and basins, sheet structures and tafoni, and several other features characteristic of granite exposures, have generated discussion in several languages (Reusch, 1883; Popoff and Kvelberg, 1937; Matschinski, 1954). Inselbergs are spectacular landscape features, so much so that they have merited a special issue of the Zeitschrift für

6   *Landforms and Geology of Granite Terrains*

Figure 1.3. Typical granite landforms: (a) Boulders at Palmer, eastern Mt Lofty Ranges, South Australia. (b) Bornhardts in central Namibia. (c) Nubbin in northwest Queensland.

Characteristics and foundations 7

Figure 1.3. (d) Castle koppie in central Zimbabwe. (e) Inselberg landscape in Namaqualand, South Africa. (f) Rock basin on Dartmoor, southwestern England.

8 Landforms and Geology of Granite Terrains

Figure 1.3. (g) Fluting at Pic Boby, Andringitra syenitic massif, Madagascar.

Geomorphologie (Bremer and Jennings, 1978). Granite forms are discussed in many general textbooks of geomorphology, as well as in regional accounts and in monographs concerned with particular aspects of geomorphology. For example, granite forms feature prominently in Worth's (1953) Dartmoor essay and in King's perceptive account of the landscapes of southern Africa, as well as in Thomas' texts (1974, 1994), dealing with tropical geomorphology, in Büdel's (1977) analysis of climatic geomorphology, and in Ollier's treatise (1969) on weathering. They are prominent in monographs concerned with structural landforms as, for example, those published by Birot (1958), by Twidale (1971, 1982) and by Ritchot (1975), as well as in the analysis of the shield lands due to various French authors and edited by Godard, Lagasquie and Lageat (1994).

There are several regional accounts and published theses of areas that are largely granitic. Granite exposures in various parts of Europe have stimulated many investigations and publications. As becomes clear in the following chapters, it is no accident that several characteristic minor granite forms were first described in the scientific literature as a result of experience in European uplands. More recently Pedraza, Sanz and Martín (1989) have provided a description of the granite forms of La Pedriza, in central Spain, and Roqué and Pallí (1991) have carried out detailed studies of the granite morphology of the Catalonian Costa Brava. Klaer's (1956) penetrating accounts of the granite landforms of Corsica are justifiably well-known, and Lagasquie (1978) has published a detailed and perceptive analysis of granite forms in parts of the Pyrenees. And despite an understandable focus on glacial and periglacial or nival geomorphology, Scandinavian workers have followed a long-standing tradition (Hult, 1873), with several distinguished granite studies. French, British and German investigators have shared this interest in cold lands Godard, Lagasquie and Lageat (1994), but have also made signal contributions to the study of granite in the tropics (Petit, 1971; Branner, 1896; Choubert, 1974).

Many postgraduate theses are based on granite morphology. For instance, in the Iberian Peninsula, the Sierra Guadarrama has provided materials for doctoral theses such as those due to Centeno (1987) and Sanz (1988); the Costa Brava has been investigated by Roqué (1993) the Galician massifs by Vidal Romaní (1983) and Uña Álvarez (1986), and northern Portugal by Coudé Gaussen (1981). Similarly, Bourne's (1974) thesis was concerned with the granite forms of northwestern Eyre Peninsula; and Campbell's (1990) Ph.D. dissertation was based on the Gawler Ranges (South Australia) which, though developed in volcanic dacite and rhyolite, is germane to granite landform studies.

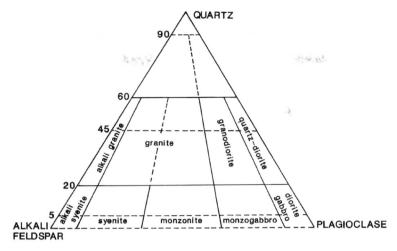

Figure 1.4. Composition and classification of common granitic rocks, according to Streckeisen (1967).

Wilhelmy (1958) published a monograph concerned specifically with granite forms, but with the emphasis on climatic influences, and Godard (1977) has provided a delightfully concise account of granitic terrains considered both from the geomorphological and geographical points of view, the last contribution is the monographic book of Ikeda (1998). Some years ago Twidale (1982) and later Vidal Romaní and Twidale (1998) published an account of granite geomorphology which in some measure forms the basis of the present review, though the scope of the present monograph is broader and the analysis is both extended and updated.

1.3 OCCURRENCES OF GRANITE

Granitic rocks, i.e. granite *sensu stricto* or its close petrological relatives (see Fig. 1.4 and Streckeisen, 1967) form the rafts of sial (a mnemonic derived from its two major components, silica and alumina) materials that underlie the continents, though, enigmatically, granite occurs in such oceanic islands as the Seychelles. Such occurrences can be explained in terms of plate tectonics, the Seychelles having splintered from the host crustal block (what is now the Indian Peninsula) and been left behind during the migration of the latter. Granites form extensive outcrops in the shield lands that form the ancient nuclei in each of the continents (Fig. 1.5), and also of orogens. They also protrude through the sediments of platform areas in small, but frequently notable, exposures.

The continental land masses occupy almost one third of the Earth's surface, though both on the continents and on the continental shelves and slopes, veneers of other materials reduce exposures of granite to about 15% of the continental areas, or about 4.5% of the Earth's surface.

Many granites originate in subduction zones where plates are melted and the resulting magma is fractionated. Where continental plates collide and obduct pre-existing acid rocks are remelted creating migmatites or hybrid magmas. Thus, many granite masses result from the repeated emplacement, magmatic assimilation and fractionation of crustal materials. They are acidic magmas which did not reach the surface and cooled comparatively slowly, allowing time for relatively large crystals to form. The larger masses of granite, and especially migmatites, however, are metamorphic. They formed as a result of the transformation of other rocks, including argillites, by hot fluids and gases spreading through the Earth's lithosphere. These processes have gone on throughout geological time, and they continue to operate.

Regardless of their origins and petrological characteristics, these magmatic bodies are referred to as plutons, after Pluto, the god of the underworld and of the dead in Greek and Roman mythology.

10 *Landforms and Geology of Granite Terrains*

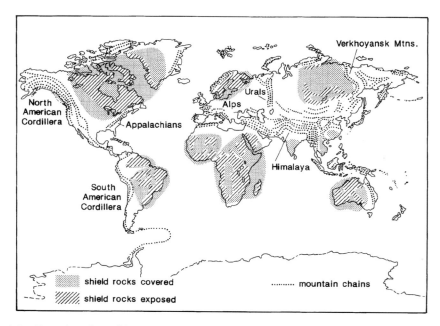

Figure 1.5. Tectonic regions of the continents: granitic rocks occur overwhelmingly in the shields and orogens.

Plutons are classified partly on their size and shape, but mainly according to whether they are predominantly discordant or concordant in respect of structures in the host mass. Batholiths are massive intrusive bodies, and are oval or shield-shaped in plan. They appear either to maintain or increase in diameter in depth, and they eventually may cut out in depth, suggesting a globular or lenticular overall form. Most batholiths are complex for they consist of several individual plutons. Thus, the Sierra Nevada batholith, in California, is about 60,000 km^2 in area and comprises perhaps 200 individual bodies emplaced over some 100 Ma. Stocks are small batholiths, conventionally less than 100 km^2. Diapirs are discordant globular bodies terminating below in the pipe or conduit through which the magma was emplaced. Laccoliths are intrusive masses located at shallow depths that have caused doming of the host rocks, phacoliths sheet-like bodies located in the crests of anticlines, lopoliths lenticular masses with sunken centres, gneiss domes, structural domes in granitic rocks, and so on. Sills are tabular bodies emplaced parallel or concordantly with such structures as bedding or foliation in the country rock, whereas dikes (or dykes) are of similar geometry but are discordant with structures in the host rock. Ring complexes are oval, circular or arcuate groups of sills or dikes related to an intrusive centre. Though they vary in shape and size, all these bodies originated deep in the crust and are therefore plutonic.

1.4 GRANITE – DEFINITION AND COMPOSITION

The word granite is derived from the Italian granulo, meaning a grain or particle, and was first used by Caesalpus in 1596 (in Twidale, 1982). From the Renaissance onwards the term was used of all crystalline rocks. Nowadays, the word granite is widely recognised in a general context, so much so that laymen commonly refer to any crystalline rock as granite. The rock itself is also familiar to the general public, for granite is a beautiful ornamental stone and, polished, is widely used as a facing on major buildings and monuments. Many headstones of graves are also of granite.

Various definitions have been suggested (Read, 1957), but granites can be taken as being plutonic rocks, in which individual crystals are a few millimetres diameter, are clearly visible to the naked eye, and are described as macrocrystalline. Granite contains at least 10% and up to 40% free

Characteristics and foundations 11

Figure 1.6. (a) Thin section showing medium grained equigranular granite texture; scale in mm.

quartz, together with feldspar and mica. Both alkali (K and Na rich, orthoclase, microcline) and plagioclase (Na and Ca rich, albite, oligoclase, andesine, etc.) feldspars are present in most granites. Relative frequencies of alkali and plagioclase feldspars, plus grain size, form the basis of subdivision and classification of granitic rocks (Fig. 1.4).

Many of the names given to granitic rocks are self explanatory, so that a granodiorite has mineralogical characteristics intermediate between a granite and a diorite; but others, still occasionally used and of historical interest, are of local derivation. Thus, another name for granodiorite is tonalite, after the Tonale Alps in northern Italy. The nearby Adamello Alps have given their name to adamellite. Syenites take their name from Syene, now known as Aswan, in Egypt, and monzonite, well known from its occurrence in the Yosemite region of the Sierra Nevada, California, from Monzoni, in the Tyrol.

Granites (and other crystalline rocks) in which the constituent crystals are completely bounded by faces are said to be euhedral. Where such faces are absent the rock is described as anhedral, and where the rock is partly euhedral, partly anhedral, it is said to be subhedral. The texture of rocks with anhedral crystals, with shapes dictated by adjacent grains, is described as allotriomorphic or xenomorphic, in contrast with euhedral or subhedral forms with a hypidiomorphic (or subidiomorphic) texture. Granites in which the constituent crystals are all of approximately the same size are described as equigranular (Fig. 1.6a). They may, however, be fine grained (e.g. aplite) or coarse (e.g. pegmatite). Some crystalline rocks display a markedly bimodal grain size with very large crystals or phenocrysts set in a finer groundmass and are described as porphyritic (Fig. 1.6b). Granites are light coloured, or leucocratic (as opposed to dark-coloured or melanocratic), commonly pink or grey in overall colour, and typically massive. Granodiorite is by far the most common of the granitic rocks, ocurrences being equal in extent to all the other types together. Granitic rocks grade into a group of crystalline rocks in which free quartz is accessory and not essential, but which are again classified on the basis of contained feldspars. These rocks are known generally as granitoids. Clearly, granitic rocks vary in composition and texture, not only between, but also within, batholithic or other plutons. For this reason alone, different parts of the same pluton may vary in their responses to weathering agencies, and hence give rise to different landform assemblages.

Crystal orientation in granitoids is generally random, with only weak and local alignment of crystal axes, as for instance in the immediate vicinity of some fractures, presumably those along

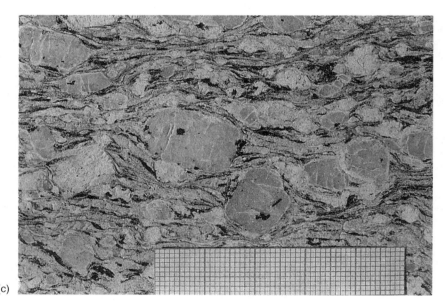

Figure 1.6. (b) Porphyritic granite, South East district of South Australia. (c) Thin section of gneissic granite with augen ("*eyes*") of potash feldspar and plagioclase set in layered matrix of biotite and quartz; scale in mm.

which there has been some dislocation. Such lineation is much more pronounced in gneisses (Fig. 1.6c), where there is, in addition, a tendency for minerals to form distinct layers, or folia, producing distinct planes of splitting or foliation (Fig. 1.7a). Planes of easy parting known as cleavage and schistosity (and also strain in individual crystals, and notably in quartz) are caused by brittle-state fracture. These features markedly influence the penetration of water into the rock mass and thus the

Characteristics and foundations 13

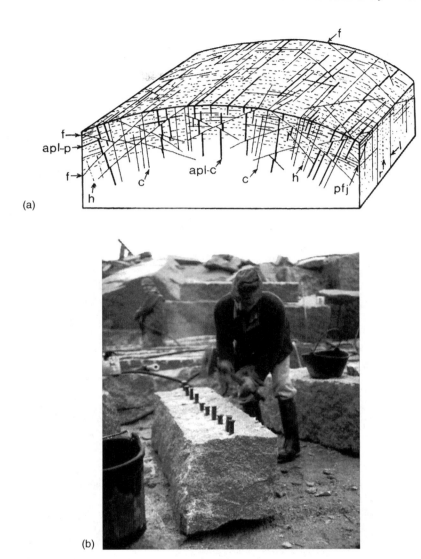

Figure 1.7. (a) H. Cloos' diagram of fractures in a batholith: c – cross joints, l – longitudinal joints, f – flat-lying joints, some of them stretching planes, r – rift, h – hardway, apl-c and apl-p are aplitic dikes (sills?), and dashes are linear flow structures. (b) Quarrymen at the Porriño Quarry, South Galicia, Spain splitting the rock along the rift.

location and course of weathering, erosion and eventual morphology. Granites of similar petrology may give rise to different landform assemblages according to tectonic history.

1.5 PHYSICAL CHARACTERISTICS

Granite has an average specific gravity of 2.662. A cubic metre of granite weighs of the order of 2,658 kg, or almost 2 tonnes a cubic yard. Its physical hardness varies according to composition, and principally according to the proportion and type of feldspars present. Despite its crystallinity,

Figure 1.8. (a) Orthogonal fractures exposed in quarry near Blackingstone Rock, eastern Dartmoor, southwestern England (b) Orthogonal fractures cut by faults exposed in quarry near Lugo, Galicia.

granite at the surface is flexible in thin sheets. Like most rocks fresh granite has a considerable compressive strength, but it also possesses a high tensile strength (Dale, 1923).

Fresh granite is of low porosity and permeability, but is pervious because near the surface it is characteristically fissured and fractured. Porosity, also known as mass permeability, refers to the ratio of the volume of voids to the total volume of rock expressed as a percentage. Porosity depends on the shape of constituent grains as well as their sorting, packing and cementation. A mass of closely packed uniform spheres consists of 26% by volume of voids or pore space, but in a crystalline medium such as fresh granite the value is commonly less than 1%. Permeability, also known as primary permeability,

Characteristics and foundations 15

Figure 1.8. (c) Vertical air photograph showing orthogonal fracture systems in granitic rocks in central Labrador (Department of Energy, Mines and Resources, Canada).

refers to the capacity of a medium to transmit fluids. It differs from porosity for voids may be unconnected, or be too narrow to allow transmission of fluids because of surface tension. Fresh granites are of low permeability but most weathered rocks allow ready passage of fluids.

Perviousness is also known as secondary or acquired permeability and refers to the capacity of a rock to transmit fluids not through the rock mass, but by way of voids, fractures and fissures. Perviousness varies not only with the number of fractures per unit volume but also with their condition – open or hairline (tight) – and according to whether they are connected.

Quarrymen working granite have long recognised that the rock splits more readily in certain directions than in others (Dale, 1923) (Fig. 1.7b). In the vertical or near-vertical plane three directions are recognised. Rift (Tarr, 1891) is the direction of easiest splitting, grain is fairly easy, and hardway the most difficult. Lift is a horizontal plane of easy splitting. Interpretations of these directions vary but many consider that rift consists of microfissures and crystal dislocations related to tectonic stress. In addition to these planes of weakness, visible partings or fractures are also well-developed. Fractures along which there is no detectable dislocation are called joints, whereas those along which there has been relative movement are faults. It is frequently difficult to prove dislocation along fractures in granite (though it is possible to demonstrate small displacements by matching crystal boundaries on opposite sides of partings), and many fractures termed joints are, it is suspected, really faults. The importance of fractures as a means of water penetration and transmission is enhanced by the low porosity and permeability of fresh granite.

Fractures or partings that run in parallel are called sets, whereas repeated patterns (which comprise combinations of different sets) are known as systems. Fracture patterns in granite are frequently

16 *Landforms and Geology of Granite Terrains*

Figure 1.9. Sheet fractures exposed in coastal cliffs, Pearson Islands, Investigator Group, eastern Great Australian Bight.

complex and apparently random or nonsystematic, though others form repeated patterns and are said to be systematic. The origin of many joints is still unclear. Explanations based on desiccation, fluid pressure, Earth's tides, and stress have been suggested.

In granite as in other rocks, many joints are nonsystematic, but one system and one set of fractures are widely developed (Kendall and Briggs, 1933; Davis, 1984). An orthogonal or, frequently, a rhomboidal system, consisting of three sets intersecting approximately at right angles to one another is characteristic of granite but also of many other massive rocks (Fig. 1.7a). The set consists of arcuate partings essentially, and most commonly, disposed parallel to the land surface and is known as sheet fracture (Fig. 1.9). Both sets or types are widely developed and they commonly co-exist. In granitic terrains they are of prime importance because of their direct impacts on landform development. In addition, their origins carry implications for the genesis of various granite forms, major and minor. For this reason these major fracture systems are discussed separately.

1.6 ORTHOGONAL FRACTURE SYSTEMS

Orthogonal systems in granite occur at various scales from the regional to the site (Figs 1.7 and 1.8b). Cloos (1923, 1931) related orthogonal systems to flow banding in batholiths (Figs 1.10a and b and Balk, 1937) has suggested that some steeply dipping fractures may be radiated or fan joints. Orthogonal or rhomboidal patterns of fractures related to shear stress are called conjugate joints, or shears if they are demonstrably faults. The fractures develop at an angle to the plane of greatest principal stress ($s1$ – Fig. 1.11), with the axis of $s1$ splitting the acute angle between the conjugate fractures. Cloos (1931) showed experimentally that orthogonal or rhomboidal systems of fractures develop as a result of compression, the conjugate shears forming two sets each disposed at roughly 45° to the direction of stress, but in reality the angular relationship between the two shear planes varies according to the force applied and the nature of the country rock. Any tension fractures caused by brittle failure develop at right angles to $s3$, the plane of least main stress. In many areas, orthogonal systems are geometrically related to regional structures, such as lineaments and known major faults, and are probably due to crustal stress. Thus, the regional orthogonal sets of Eyre Peninsula (South Australia) and the adjacent Gawler Ranges (Hills, 1956; Twidale and Campbell, 1990) are disposed at 45–60° to ancient lineaments, such as the Hinge Zone or Torrens Lineament

Characteristics and foundations 17

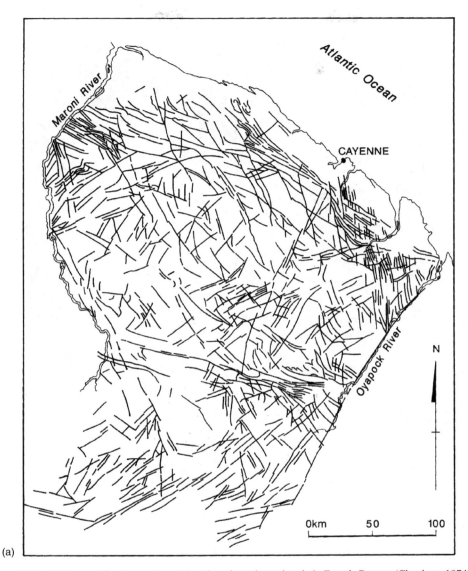

Figure 1.10. Orthogonal fracture patterns (a) at the subcontinental scale in French Guyana (Choubert, 1974).

and/or the Lincoln Fault but in parallel with the Australian lineament pattern. A similar relationship is discernible in French Guyana (Choubert, 1974) (Fig. 1.10a) where the major fractures trend NNW-SSE and WNW-ESE, with the minor systems disposed either in parallel or NW-SE and NE-SW. Such patterns have led some to consider that orthogonal fracture patterns originate as a result of crustal stresses developed either during the emplacement of the crystalline masses, or imposed subsequently. Many consider they have developed in relation to horizontal shearing in the crust, such as that associated with plate migrations, past and present. This suggestion is sustained by the swarms of minor fractures formed in parallel with the main partings. At a local scale, some clefts or slots are linear, and run parallel to observed fractures, but are not themselves associated with partings or at least not throughout their length.

18 *Landforms and Geology of Granite Terrains*

Figure 1.10. (b) At the local scale, as exposed on the surface of Pildappa Rock, northwestern Eyre Peninsula, South Australia.

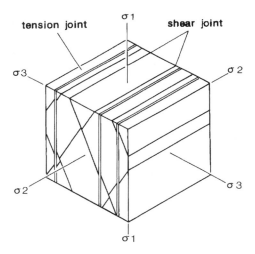

Figure 1.11. Orthogonal fracture system and stress directions (Davis, 1984).

1.7 FRACTURES AND DRAINAGE PATTERNS

Most drainage patterns are determined by the structural characteristics of the country rock and the slope of the land surface (Zernitz, 1931). On gentle slopes or plains, structural control is minimal and parallel, subparallel or dendritic patterns are developed. Thus, the regional drainage pattern on Dartmoor, southwestern England, is radial, reflecting the domical topography of the region (Fig. 1.12a). Quite commonly in granitic terrains, systems of steeply inclined orthogonal or rhomboidal fractures have been exploited by rivers to give straight, angular or rectangular stream patterns, according to the precise geometry of the fracture systems (Figs 1.12b and 1.13a and b), and even where slope is the dominant factor, angular patterns are developed locally in response to structure (Fig. 1.12c). The reason for the coincidence between fracture and stream channel is that fractures

Characteristics and foundations 19

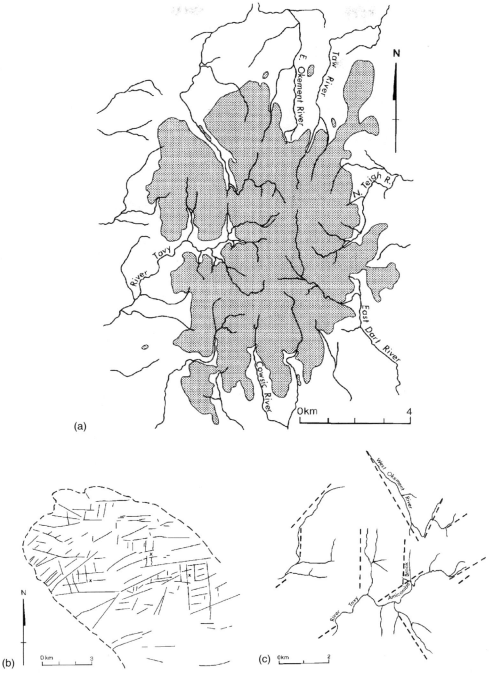

Figure 1.12. (a) Radial drainage pattern on Dartmoor, southwestern England. (b) Rectangular stream pattern (especially in areas marked X) in granitic exposure in the Cairns-Mosman area, north Queensland. (c) Local rectangular drainage pattern on part of northwestern Dartmoor, England. Suggested fracture trends indicated by dashed lines.

20   *Landforms and Geology of Granite Terrains*

(a)

Figure 1.13.   (a) Drainage patterns in granitic terrains: straight, fracture-controlled valley near Kylie Lake, southern Zimbabwe.

are zones of weakness which can readily be penetrated by meteoric waters. The bedrock adjacent to the fracture zones is weathered and is thus susceptible to erosion by streams, which become the prominent members of the drainage network by natural selection; for their competitors draining unweathered granite do not extend or incise as rapidly as do those which run along fractures.

In granite as in other lithological and structural environments, anomalous drainage patterns are developed in places, i.e. river and stream patterns do not conform to lines of structural weakness in the country rock, but rather run transversely to the structural grain, local or regional. As in other places recourse must be made to diversion, antecedence, superimposition, inheritance, stream persistence and valley impression in explanation of such transverse drainage.

Thus, the Vaal River is anomalous with respect to bedrock structure where it flows over the northern edge of the Vredefort Dome (King, 1942), including part of the granite core, in the Free

Figure 1.13. (b) Fault zones exploited by weathering and erosion, Coast Mountains, British Columbia (Department of Lands, Forests and Water Resources, British Columbia).

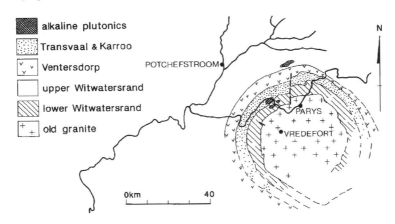

Figure 1.14. The Vaal River crossing the Vredefort Dome at Parys, South Africa (du Toit, 1954).

State, South Africa (Fig. 1.14). The river must have developed at least the main features of its present course on a much higher surface prior to encountering the Vredefort structure.

According to King (1942) it is superimposed from Karroo (late Palaeozoic-Early Mesozoic) strata which once completely overlay the region. Even so, the high discharge of the Vaal and positive

22 Landforms and Geology of Granite Terrains

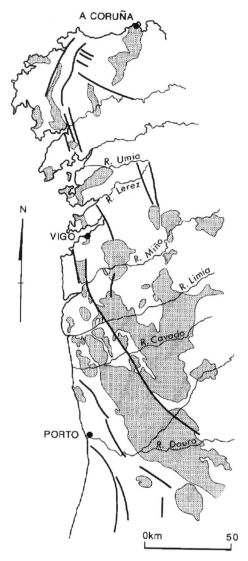

Figure 1.15. Transverse stream patterns on the Atlantic slope of the northwestern Iberian Peninsula (Adapted from Parga Pondal et al., 1982).

feedback or reinforcement mechanisms have together enabled it to maintain its course across the strike ridges of the Witwatersrand system, as well as the structures of the granitic core at the local scale, where straight stream sectors reflect fracture control. Superimposed, like antecedent, rivers imply unequal activity and the capacity of rivers to maintain their courses across transverse structures.

The Western Iberian Peninsula drainage provides especially clear examples. Structurally the region is dominated by Variscan orogenic trends running NNW-SSE (Iglesias and Choukrounne, 1980; Parga, Parga, Vegas et al., 1982). The many granite batholiths of the area intruded the previous sedimentary and metamorphic terranes which are more susceptible to weathering and erosion than the granitic rocks. Nevertheless, several of the rivers flowing to the Atlantic, and notably the Xallas, Tambre, Ulla, Umia, Miño, Lima, Cávado and Douro, flow across the structural grain and across some granite emplacements (Fig. 1.15 and see Richthofen, 1886). These rivers drain the inner plains of Iberia inherited from the Pangaean stage. In the northwestern Iberia there are preserved

Characteristics and foundations 23

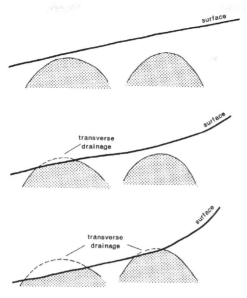

Figure 1.16. Diagrammatic cross-section showing inherited drainage being impressed on granite batholiths during incision.

Figure 1.17. Waterfalls in Río Xallas, Lézaro, Galicia, Spain.

remnants of these inner plains that allow defining the Fundamental Surface of Galicia (Vidal Romaní, Yepes and Rodríguez, 1998). The Iberian atlantic rivers, as its cutdown, are so later to the Atlantic Ocean opening that started about the Jurassic times. From that time the fluvial net incision is the answer to the creation of the new marine base level to the thining of the litosphere affected by the system of listric faults and the slight isostatic rebound typical of the continental plate passive margins. The incision process has not yet finished totally as it may be proved in the longitudinal profiles of all the Iberian atlantic rivers (Fig. 1.16), and it is especially evident in the case of the Xallas river, which meets the sea in a waterfall of 30 m of height (Vidal Romaní, Yepes and Rodríguez, 1998), (Fig. 1.17).

REFERENCES

Balk, R. 1937. Structural behaviour of igneous rocks. Geological Society of America Memoir 5.
Birot, P. 1958. Morphologie Structurale. Volume 1. Presses Universitaires de France, Paris.
Bourne, J.A. 1974. Chronology of denudation of northern Eyre Peninsula. Unpublished M.A. thesis, University of Adelaide.
Branner, J.C. 1896. Decomposition of rocks in Brazil. Geological Society of America Bulletin 7, 255–314.
Bremer, H. and Jennings, J.N. (eds.) 1978. Inselbergs/Inselberge. *Zeitschrift für Geomorphologie* Supplement Band 31.
Büdel, J. 1977. Klima-Geomorphologie. Borntraeger, Berlin.
Campbell, E.M. 1990. Structure and surface in the Gawler Ranges, South Australia. Unpublished Ph.D. thesis, University of Adelaide.
Centeno, J.D. 1987. Morfología granítica de un sector del Guadarrama Oriental (Sistema Central Español). Tesis Doctoral, Universidad Complutense de Madrid.
Choubert, B. 1974. Le Précambrien de Guyanes. Memoir Bureau de Recherches Géologiques et Minières 81.
Cloos, H. 1923. Das Batholithenproblem. *Fortschrift der Geologie und Palaeontologie* 1: 1–8.
Cloos, H. 1931. Zur experimentellen Tektonik. Brüche und Faltung. *Die Naturwissenschaften* 242–247.
Coudé Gaussen, G. 1981. Les Serras de Peneda et do Gêres. Etude géomorphologique. Universidade de Lisboa. Instituto Nacional de Investigação Cientifica. Memorias do Centro de Estudos Geográficos, Lisbon.
Dale, T.N. 1923. The commercial granites of New England. United States Geological Survey Bulletin 738.
Davis, G.H. 1984. Structural Geology of Rock and Regions. Wiley, New York.
Godard, A. 1977. Pays et paysages du granite. Presses Universitaires de France. Paris.
Godard, A., Lagasquie, J.J. and Lageat, Y. 1994. Les regions de Socle. Faculté des Lettres et Sciences Humaines de l'Université Blaise-Pascal Nouvelle Serie, Fascicule 43.
Hills, E.S. 1956. A contribution to the morphotectonics of Australia. *Journal of the Geological Society of Australia* 3: 1–15.
Hult, R. 1873. Fran Nord till Syd Kalender Fjfillvandringar i Galicien i och Zamora. *Geografiska Foreningen i Finland* 30–55.
Iglesias, M. and Choukrounne, P. 1980. Memoria y Hoja n° 228 (Viana del Bollo). Mapa Geológico de España. Instituto Tecnológico Geominero de España. Madrid.
Ikeda, I. 1998. The world of granite landforms. Kokon-Shoin. Tokyo, Japan.
Kendall, P.F. and Briggs, H. 1933. The formation of rock joints and the cleat of coal. *Proceedings of the Royal Society of Edinburgh* 53: 167–187.
King, L.C. 1942. South African Scenery. Oliver and Boyd, Edinburgh.
Klaer, W. 1956. Verwitterungsformen in granit auf Korsika. Petermanns Geographischen Mitteilungen Ergännzungsheft.
Lagasquie, J.J. 1984. Géomorphologie des Granites. Les Massifs Granitiques de la Moitié Orientale des Pyrenées Françaises. Editions de le Centre Nationale de la Recherche Scientifique.
Matschinski, M. 1954. Quelques considérations sur la théorie mathématique des taffoni. Accademia Nazionale dei Lincei, Atti. *Classe de Scienze Fisiche Matematiche e Naturali Rendiconte* (Series 8) 16: 632–633, 731–734.
Ollier, C.D. 1969. Weathering. Oliver and Boyd, Edinburgh.
Parga, I., Parga, J.R., Vegas, R. and Marcos, A. 1982. Mapa Geológico del Macizo Hespérico 1: 500 000. Edicións O Castro. Coruña, Spain.
Pedraza, J., Sanz, M.A. and Martín, A. 1989. Formas graníticas de La Pedriza. Cuadernos madrileños del Medio Ambiente. Agencia de Medio Ambiente de la Comunidad de Madrid. Madrid.
Petit, M. 1971. Contribution à l'etude morphologique des reliefs granitiques à Madagascar. Société Nouvelle de l'Imprimerie Centrale, Tananarive.
Popoff, B. and Kvelberg, I. 1937. Die Tafoni-Verwitterungserscheinung. Latvijas Universitates Raksti. *Acta Universitatis Latviensis*. Kimijas Facultates Serija IV. 6: 129–368.
Read, H.H. 1957. The Granite Controversy. Murby, London.
Reusch, H.H. 1883. Note sur la géologie de la Corse. Paris. *Societé Geologique de France Bulletin* 11: 53–67.
Richthofen, F. von 1886. Führer für Forschungsreisende. Jenecke, Hanover.
Ritchot, G. 1975. Essais de Géomorphologie Structurale. Presses Universitaires Laval, Québec.
Roqué, C. 1993. Litomorfologia dels Massissos de Les Gavarres i de Begur. Tesis Doctoral. Universitat Autònoma de Barcelona, España.
Roqué, C. and Pallí, Ll. 1991. Modelat del Massís de Begur. *Estudis del Baix Empordá*, 10: 5–48. Girona, España.

Sanz, C. 1988. El Relieve del Guadarrama Oriental. Comunidad de Madrid. Consejería de Política Territorial. Madrid, Spain.
Streckeisen, A.L. 1967. Classification and nomenclature of plutonic rocks. *Geologische Rundschau* 63: 773–786.
Tarr, R.S. 1891. The phenomenon of rifting in granite. *American Journal of Science* 41: 267–272.
Thomas, M.F. 1974. Tropical Geomorphology. Macmillan, London.
Thomas, M.F. 1994. Geomorphology in the Tropics. Wiley, Chichester.
Toit, A.L. du 1954. The Geology of South Africa (Third Edition, revised). Oliver and Boyd, Edinburgh.
Twidale, C.R. 1971. Structural Landforms. Australian National University Press, Canberra.
Twidale, C.R. 1982. Granite Landforms. Elsevier, Amsterdam.
Twidale, C.R. and Campbell, E.M. 1990. Les Gawler Ranges, Australie du Sud: un massif de roches volcaniques silicieuses, à la morphologie originale. *Revue de Géomorphologie Dynamique* 39: 97–113.
Uña Álvarez, E. de 1986. El Macizo de A Coruña. Análisis estructural y morfología de un afloramiento granítico. Tesis Doctoral. Universidad de Santiago de Compostela.
Vidal Romaní, J.R. 1983. El Cuaternario de la provincia de La Coruña. Modelos elásticos para formación de cavidades. Tesis Doctoral, Universidad Complutense de Madrid. Publicaciones de la Universidad Complutense de Madrid, Madrid.
Vidal Romaní, J.R. and Twidale, C.R., 1998. Formas y Paisajes Graníticos. Serie Monografías 55, Universidad de A Coruña, Servicio de Publicaciones. A Coruña.
Vidal Romaní, J.R., Yepes, J. and Rodríguez, R. 1998. Geomorphic evolution of the Peninsular Hesperian Massif. Study of a sector situated between Lugo and Ourense Provinces (Galicia, NW Spain). *Cadernos do Laboratorio Xeolóxico de Laxe* 25: 165–199.
Wilhelmy, H. 1958. Klimamorphologie der Massengesteine. Westermanns, Bruswick.
Worth, R.H. 1953. Worth's Dartmoor In Spooner, G.M. and Russell, R.S. (eds). David and Charles, Newton Abbott.
Zernitz, E.M. 1931. Drainage patterns and their significance. *Journal of Geology* 40: 498–521.

2

Sheet fractures and structures

2.1 TERMINOLOGY

Many granite outcrops are subdivided not only by orthogonal or rhomboidal sets of fractures but also by flat-lying or gently arcuate partings (Figs 1.8 and 2.1). The latter are of two types. One set is surficial partings, and is referred to as pseudobedding, pseudostratification, or flaggy joints. They are essentially discontinuous, occur within a few metres of the surface and affect morphology at a small scale, banks of shallow clefts being produced by preferential weathering of the partings (Fig. 2.2).

Sets of horizontal or curvilinear fractures that extend to greater depths than pseudobedding are known by various names: flat-lying joints, Lägerklufte, Bankung, structure en gros bancs, estructura en capas, stretching planes, shells, and exfoliation, or offloading, relief of load, pressure release, sheeting or sheet joints. For various reasons, but mainly because they preempt discussion of origin, several of these terms are unsuitable, and here sheet, or sheeting fractures, is preferred as genetically neutral but yet descriptive; for though the word sheet may suggest a thin layer, whereas some of the forms discussed here are 10 m or more thick, they are nevertheless thin in the global, continental or regional contexts. Fracture is preferred to joint because dislocation is evident along some

Figure 2.1. Sheet fractures (a) in granite in the Porriño, Budiño, Galicia, Spain.

28 *Landforms and Geology of Granite Terrains*

Figure 2.1. Sheet structure (b) exposed in granite hill in the Pindo, Galicia, Spain, and showing increase in radius of sheet fractures with depth; (c) on granite bornhardt in the western Sahara, Mauritania (Peel, 1941).

of the partings, which are, therefore, small displacement faults. Sheet structure is used to denote the massive slabs defined by the sheeting fractures.

2.2 DESCRIPTION AND CHARACTERISTICS

Sheet structure consists of thick arcuate slabs defined by sheet fractures (Figs 2.1 and 2.3). The slabs are up to 10 m thick, and are arbitrarily defined as being more than about 0.2 m thick (slabs

Figure 2.2. (a) Pseudobedding at Roughtor, Bodmin Moor, southwest England (Geological Survey Museum UK). (b) Pseudobedding in monzonite, the Yosemite, Sierra Nevada, California.

less than that are termed spalls or flakes). Sheet fractures have been observed at depths of 100 m or more in some quarries, though elsewhere they fade with depth. Indeed, in some areas with well-developed orthogonal fracture systems, as in some of the residuals found on northwestern Eyre Peninsula, and in the silicic volcanics of the Gawler Ranges, in the arid interior of South Australia, sheet fractures appear to be superficial (Campbell and Twidale, 1991). On the other hand, it is evident in some deep mines and other excavations that sheet fractures extend to great depths. Some sheet fractures take the form of simple arcuate partings. Others consist of several separate fractures arranged en echelon and together forming an arcuate parting (Fig. 2.3a). It is frequently claimed that the thickness of sheet structure increases systematically with depth, but there are many exceptions.

30 *Landforms and Geology of Granite Terrains*

Figure 2.2. (c) Pseudobedding cutting across orthogonal fractures, at Heltor, Dartmoor, southwest England.

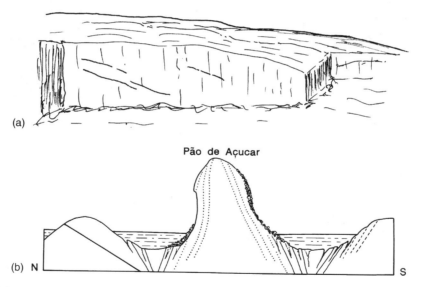

Figure 2.3. (a) Sketch of quarry face near Wudinna, northwestern Eyre Peninsula, South Australia, showing en echelon arrangement of partings in complex sheet fracture. (b) Geological section through the Pão de Açucar, Rio de Janeiro, Brazil.

At some sites radius of curvature increases with depth (Fig. 2.1b). Most, though significantly not all (see below), sheet fractures run roughly parallel to the land surface, being essentially horizontal on summits and beneath valleys, but steeply inclined beneath hillsides (Figs 2.1 and 2.3b). At many sites, however, the sheeting fractures are more steeply inclined than the land surface (Fig. 2.3c) while at others the hillslope truncates the partings (Fig. 2.3d). They also plunge steeply (up to 70°) in the vicinity of major vertical or subvertical fractures, and some terminate against such partings (Fig. 2.3e). Many so-called sheets are attenuated wedges, though it is in some instances difficult to distinguish between original geometry and modifications due to weathering and

Sheet fractures and structures 31

Figure 2.3 (c) View of White's Quarry, near Minnipa, northwestern Eyre Peninsula, South Australia, showing inclination of sheet fractures greater than the slope of the hill. Note also that two structural domes (X and Y) are contained within the one topographic residual. (d) Sketch of monzonite hill in the Yosemite, Sierra Nevada, California, showing hillslopes truncating sheeting. (e) Termination of sheet fractures against steeply inclined fracture, north side of Mariz Quarry, Galicia, northwestern Spain.

Figure 2.3. (f) Wedge-shaped sheet structure at northern end of Ucontitchie Hill, northwestern Eyre Peninsula, South Australia: the attenuation is, almost certainly and in large part, due to weathering and erosion.

Figure 2.4. Discontinuous sheet fractures in a Brazilian residual (Branner, 1896, and Lamego, 1938).

erosion (Fig. 2.3f). Moreover, though some sheet structures are continuous features (Fig. 2.1b), others appear to fade and disappear toward the interior of the host mass (Fig. 2.4).

Sheet fractures are very well and widely developed in granitic rocks, including gneiss and migmatite, but are also found in other massive rocks such as dacite and rhyolite, sandstone, conglomerate and limestone (Chapter 12). Sheeting planes cut across other bedrock structures including orthogonal systems, columnar joints, cleavage and foliation, crystal boundaries, rift and grain, flow structures and bedding. Though some sheet structures are recent developments, having formed in relation to youthful surfaces shaped by riverine or by glacial erosion, elsewhere they are of some antiquity. On Dartmoor, England, for example sheeting planes are intruded by Mesozoic sills. But sheet fractures clearly postdate the consolidation of the rock in which they occur; they are brittle fractures. As it is apparent from the examples cited and discussed, sheet fractures occur in a wide range of climatic regimes.

2.3 THEORIES OF ORIGIN

Bornhardts are invariably associated with sheet structure but whether these arcuate fractures give rise to the domed shape, or whether they are induced by it, is debatable. Two diametrically opposed views of the relationship between the form of the land surface and the geometry of sheeting joints have evolved over the past 150 years or so.

Some interpret the joints and associated sheet structure as a primary feature of the rock which has closely determined the gross morphology of the land surface. According to this view the joints were first developed in the bedrock and the shape of the land surface is a response to this internal structure. As Merrill (1897, p. 245) put it:

> "... with many geologists these joints, in themselves, would be accepted as due to atmospheric action. In the writer's opinion they are, however, the result of torsional stress and once existing are lines of weakness which become more and more pronounced as weathering progresses."

According to Merrill (1897, p. 245), the boss or dome-like form of the bornhardts is *"incidental and consequent"* on internal structure. The earliest proponent of this general endogenetic interpretation was de la Beche (see Adams, 1954) but several others, besides Merrill (1897), subscribed to the theory during the Nineteenth Century. Nowadays, however, such endogenetic theories of sheet structure development find little support, for, since Gilbert (1904) published his seminal paper on sheet fractures, the interpretation of sheet fractures as a response to the form of the land surface (Chapman, 1956) has been the most widely, indeed almost universally, accepted, as epitomised in the general usage of the terms offloading joint and pressure release joint; but wide acceptance does not necessarily imply validity.

These are the two major competing interpretations of sheet fractures and associated sheet structure, but over the years many explanations and mechanisms have been proposed. Though most fail as general explanations, some may have local validity. All fall into one of two major categories – exogenetic or endogenetic.

2.3.1 *Exogenetic explanations*

Insolation was long ago suggested as a possible cause of sheet fractures. As rocks are poor conductors of heat it has been argued that solar radiation heats the outer exposed zones of rock which expand and become detached from the main mass, forming more-or-less thick slabs or sheets. But because the effect of the Sun's radiation penetrates only a few centimetres at most into the rock, whereas sheet jointing extends to considerable depths, this view can be discounted.

- Chemical weathering has also been cited. The gradual infiltration and penetration of meteoric waters into rocks near the land surface has frequently been called upon in explanation of the flaking and spalling of rock masses. Where the chemical alteration of rocks results in increased volume and, hence, pressure, this appears feasible. However, not all chemical alteration leads to volume increase and, hence, to expansive pressure and rupture. Also, if the weathering were held to precede and to give rise to the fracturing, it is relevant to ask why chemical attack is concentrated upon, and restricted to, a few gently arcuate planes.

   Furthermore, many of the massive slabs and wedges of rock involved in sheet structure display no sign of chemical alteration. Some do (Fig. 2.5), but such alteration of minerals can more readily be explained as weathering associated with moisture seeping along pre-existing fractures, rather than weathering having caused the development of the partings.
- The suggestion that sheet fractures are an expression of offloading or pressure release is widely accepted. And all rock fractures are an expression of erosional offloading in the sense that at depth other stresses are subordinate to the pressures exerted by the superincumbent load. It is only through the release of vertical pressure that the other stresses are manifested as obvious fractures. But, a basically different interpretation of sheeting joints, which attributes them solely and wholly to pressure release without the previous application of stress, has long found favour.

The gist of the pressure release, or erosional offloading, hypothesis is that rocks which cool and solidify deep in the Earth's crust (for example granites, whether of metasomatic or igneous origin) do so under conditions of high lithostatic pressure, i.e. loading by overlying and adjacent rock. That there are widespread granite outcrops is itself proof of deep erosion, for though some plutons appear to have been emplaced at shallow depths (as for instance in eastern Papua New Guinea,

Figure 2.5. Weathering along sheeting plane (X) at Whites Quarry, near Minnipa, northwestern Eyre Peninsula, South Australia.

Ollier and Pain, 1980) most were put in position at depths of a few kilometres. Decompression achieved through the removal of superincumbent load is said to cause the development of radial stress which is tensional and is relieved by the development of fractures tangential to the stress and parallel to the land surface; these are, according to the proponents of this hypothesis, sheeting joints. The fundamental premise of the hypothesis is that the form of the land surface in broad terms determines the geometry of the sheet jointing, for it is in relation to this that the radial stress develops. Sheeting is a secondary feature formed after the development of the topography.

The general parallelism of sheeting joints and land surface can be taken as lending support to the offloading hypothesis, though the interpretation can be reversed with equally satisfactory logic. The formation of relatively thin slabs or sheets of rock close to the land surface and developed in response to recent erosion has also been accepted as persuasive evidence of offloading. For example, sheets of rock have been developed in cirque headwalls, in the floors of recently deglaciated valleys, and in relation to older glacial forms (as for instance in northern Italy), and recently dissected riverine valleys (as on the Atlantic coast of Galicia – Fig. 2.6).

Though it is plausible and persuasive, the offloading theory in the sense outlined by Gilbert (1904) and adopted by many later workers, namely, that pressure release is the sole cause of sheet jointing, may be called to question on several grounds.

– Simple triaxial tests show that compression and decompression of essentially isotropic materials do not cause fracturing save in special circumstances which are unlikely to be found in nature; such fractures are unlikely to develop in the context of slow erosional unloading (Fell, MacGregor and Stapledon, 1992). Even in anisotropic materials it appears that several cycles of compression and decompression can be applied to unconfined specimens before, with increased loads, the material ruptures in fatigue. Unloading appears to be mechanically incapable of producing sheet fractures.
– It is difficult to understand why, if expansive stress developed during erosion, it has not been accommodated along pre-existing lines of weakness. Sheeting is either absent or only poorly developed in closely-jointed granites, and as White (1946, p. 5) stated "most geologists accept

Sheet fractures and structures 35

Figure 2.6. Sheeting in granite in parallel with recently formed erosional surfaces in a deep river valley in the Pindo, Galicia; note contrast with fracture pattern and orientation exposed on upper hillslope.

this fact as evidence that the force of expansion in the rock has been dissipated by slight movements along the joint planes". In some areas, orthogonal joints predate sheet structure, but even where sheet fractures cannot be shown to be younger than the orthogonal system there are usually many other potential slippage planes in rocks and crystals. Strain could be taken up by grain boundary sliding, and along crystal cleavages, for example.

– The association of sheet structure with bornhardts is irrational if the former are interpreted as a consequence of offloading without the application of compressional stress. Sheet structure is supposed to be a manifestation of radial expansion, whereas the field evidence, both in gross and in detail, suggests that the residuals are masses of rock in compression (below, and Chapter 6). This is indicated by the condition of the joints within the bornhardts, which are tight, and commonly take the form of discontinuous hairline cracks, and by the presence of A-tents and wedges, which are clearly associated with compressional stress (Chapter 11). Many bornhardts persist as long term landforms because the rocks of which they are constructed are in compression: fractures are few and tight, water cannot readily penetrate the mass, and weathering and hence erosion are slow. If the bornhardts were indeed developed on masses of granite that were decompressed and relaxed, their joints would be open. The rock masses would not survive weathering and erosion and would not, therefore, persist as the residuals that are such prominent, and in places dramatic, landscape features.

– Although it is conceivable that the parallelism between sheet jointing and land surface need not be perfect, it is difficult in terms of the offloading hypothesis to explain inverse relationships such as have been observed in the Yosemite Valley in the vicinity of Tenaya Lake, in Joshua Tree National Monument in southern California and at Quarry Hill, near Wudinna, on northwestern Eyre Peninsula, South Australia (Figs 2.7a and b) (Twidale, 1964, 1973). Such relationships are probably the result of deep erosion and inversion of relief or of exposure of lenticular sheets distorted by faulting. If sheet fractures predated and determined the shaping of the land surface, stress contrasts alone could have produced topographic inversion through the preferential weathering of antiformal crests, leaving the synforms upstanding (Figs 2.7 (c, d, e and f) and 2.8).

– Many local and detailed lines of evidence argue against the offloading hypothesis. For instance, there is an inconsistency between the age of erosional features said to be the cause of sheeting

36 *Landforms and Geology of Granite Terrains*

Figure 2.7. Synformal structures in granite (a) at Tenaya Lake, Yosemite, Sierra Nevada, California, (b) in the Joshua Tree National Monument, southern California.

joints and the inferred age of the joints on Dartmoor, England: the former are geologically youthful whereas the latter are of considerable antiquity, so that it seems at least as reasonable to suggest that the joints determined the form of the land surface as the converse.
- The examples of fractures developed parallel to young erosional surfaces (e.g. Fig. 2.6) are better explained in terms of stress trajectories adjusting to new surface configurations, which are surfaces of least principal stress.
- Several morphological and structural features developed on and in granitic rocks are incompatible with the tensional or expansive conditions implied by offloading. For example, structural domes, wedges, and overthrusting like that developed and exposed at the Mariz quarries, near

Sheet fractures and structures 37

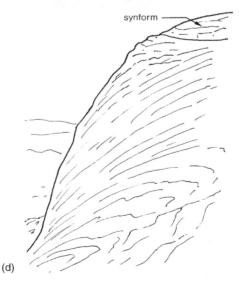

Figure 2.7. (c and d) in the headwall of a cirque in Little Shuteye Pass, Sierra Nevada, California: note the synform near the crest of the hill. See explanatory sketch and photo.

38 *Landforms and Geology of Granite Terrains*

Figure 2.7. (e and f) exposed in excavation at Quarry Hill, Wudinna, northwestern Eyre Peninsula, South Australia.

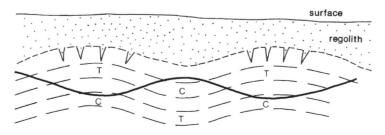

Figure 2.8. Development of bornhardts on synformal structures as a result of deep erosion and inversion of relief.

Sheet fractures and structures 39

Figure 2.9. (a) Sheet structures. (b1 and b2) Wedge in granite at Mariz Quarry, Guitiriz, Galicia.

Guitiriz, in Galicia and on Ucontitchie Hill, northwestern Eyre Peninsula, South Australia (Fig. 2.9), are impossible to explain in terms of a tensional regime, as are A-tents (Chapter 11). Evidence of dislocation along sheet fractures (e.g. Fig. 2.10) points to their being faults possibly of the bedding plane type (Vidal Romaní, Twidale, Campbell and Centeno, 1995; Twidale, Vidal Romaní, Campbell and Centeno, 1996). In some areas, e.g. Rock of Ages Quarry, Barre, Vermont – (see Fig. 2.11a), in the Rio de Janeiro area, Brazil – (Fig. 2.3b) and at sites in Galicia – (Fig. 2.11b), sheet fractures and faults coexist, suggesting that the former may be secondary shears.

Thus, there are many lines of argument and evidence which, together, strongly suggest that the offloading hypothesis cannot be uncritically accepted as an explanation of sheet jointing. Of these several considerations, undoubtedly the most significant from a geomorphological point of view is that bornhardts are rock masses in compression, whereas radial expansion and tensional stress are implied by the offloading hypothesis.

40 *Landforms and Geology of Granite Terrains*

Figure 2.9. (c) Overthrusting in granite at Mariz Quarry, Guitiriz, Galicia. (d) Wedge on the eastern side of Ucontitchie Hill, northwestern Eyre Peninsula, South Australia.

Figure 2.10. Dislocation along sheeting plane (a) in valley of Río Tabalón, Sierra Guadarrama, central Spain; (b) in quarry at La Clarté, Brittany, France, with fault steps well-developed.

2.3.2 *Endogenetic explanations*

– Turning to endogenetic explanations, several writers, including some of the earliest to consider the problems of sheet jointing, related sheet structure to the stresses imposed on magmas during injection or emplacement and, hence, to the shape of the original pluton. The parallelism of sheeting joints and the margins of the igneous body on Dartmoor were noted early last century. Some workers attributed sheet structure to a combination of stresses developed during emplacement of the granite mass and later cooling. Thus, although it may apply at a few sites, this suggestion

Figure 2.10. (c) Fault steps (X) developed on a sheet fracture exposed on the eastern side of Ucontitchie Hill, Eyre Peninsula, South Australia.

cannot stand as a general hypothesis, for inselbergs and associated sheet jointing are well-developed in rocks such as sedimentary and volcanic sequences, which have not been emplaced and even in granites the magnetic structures contemporaneous with the emplacement are clearly discordant to the sheet structure.

— Some early workers suggested that the nodular form or concentric structure of many crystalline masses was responsible for the dome-like shape of Blackingstone Rock, a tor (or inselberg) on eastern Dartmoor. The suggestion is undermined by the contrast between the northern slope of the residual, which is convex upwards and is associated with arcuate-upward fractures, and the southern, which is dominated by orthogonal fractures and has the appearance of a koppie (Figs 2.12a and b). It has been suggested that the domed inselbergs or morros of southeastern Brazil (Brajnikov, 1953; Lamego, 1938) are gigantic floaters of solid rock developed as foci of compression consequent on volume changes during metasomatism. There is, however, no reason why decompression should cause fracturing. Also, sheet jointing occurs in rocks which have not been metamorphosed.

— Because many granite masses are areas of distinct negative gravity anomaly, it has been suggested that these masses tend to rise as diapirs through the superincumbent rocks to form gneiss domes. Gneiss domes are structures developed in migmatised rocks, that is, in granitic rocks consisting partly of igneous, partly of metamorphic, materials. Foliation is well-developed and forms concentric patterns with quaquaversal dips. Gneiss domes have been described from many areas, and prominent circular structures in gneissic rocks occur in such areas as Finland, French Guyana (Choubert, 1974), Zimbabwe, the Laurentian Shield and North Carolina (Fig. 2.13). Some workers interpret gneiss domes as due to repeated injections of magma during separate orogenies or distinct phases of the same orogeny. Others consider that the structures can be explained in terms of a single phase of compression resulting in upward migration of migmatitic material to form a dome which then spreads laterally to form a mushroom-shaped mass. Whatever interpretation is placed on the structures in detail, vertical upthrust of granitic

Sheet fractures and structures 43

Figure 2.11. Sheet fractures associated with faults and overthrusting planes ("f") exposed (a) in the Rock of Ages Quarry, Barre, Vermont, in November, 1965. (b) O Pindo, Galicia, Spain.

    material is involved. In such structures, sheet jointing is construed as due to vertical movement and the development of radial stress, rather than to lateral compression; sheet jointing is caused by radial stretching introduced during uplift. A number of objections can be levelled against the hypothesis as a general explanation. For example, if denser material were displaced, then the intruded mass would surely be in compression. Sheet structure occurs in sedimentary and volcanic rocks which have not been subjected to doming, and, if the sheeting joints are stretching planes, it is difficult to explain the preservation of inselbergs as well as the other field evidence indicative of compression.
- It has also been suggested that sheeting joints may be an expression of lateral compression which results not only in faulting, but also in shearing. After the erosional removal of superincumbent

44 *Landforms and Geology of Granite Terrains*

Figure 2.12. Blackingstone Rock, eastern Dartmoor, England, seen (a) from the northwest and (b) from the south.

load, these cause the opening of previous fractures or joints of arcuate geometry, to quote Dale (1923, p. 35), as "*a series of undulating fractures extending entirely across*" a rock mass. Such fracture geometry finds many structural and topographic expressions, as for example in the rolling granite hills of the Guitiriz area of northwestern Galicia, and as exposed in many of the quarries of the district.

Sheet fractures and structures 45

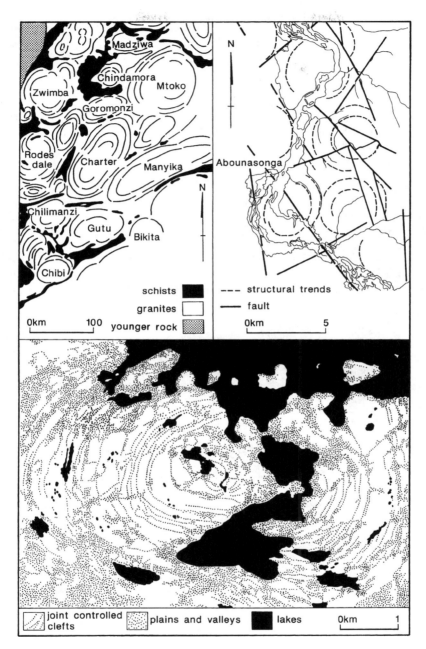

Figure 2.13.   Gneiss domes in (a) Zimbabwe (top left), (b) French Guyana (top right) and (c) Quebec, Canada (bottom). McGregor, 1951; Choubert, 1974; and air photographs of the Mingan-Natashquan area, Québec, Canada (Department of Energy, Mines and Resources, Ottawa).

Lamego (1938) clearly envisaged a similar situation obtaining in the Rio de Janeiro region, and Gilbert (1904), the author of the pressure release hypothesis, is quoted by Dale (1923) as suggesting that Stone Mountain in Georgia was due to compressive strains. Several other geologists who examined granitic domes in the northeastern United States deduced that the granites involved are in compression. Similar conclusions were reached with respect to a granite exposed in a quarry at Quenast in Belgium. In some areas, e.g. New England, there is evidence of what Dale (1923) called double-sheet structure, that is, two sets of sheet joints, the strikes of which intersect at oblique angles. This, surely, is explicable only in terms of shortening of the massif.

It has been shown by *in situ* stress measurements that the Australian continent is in a state of substantial horizontal compression, and many other examples of pronounced compressive stress in the horizontal plane, stress far greater than suggested by theoretical considerations, have been recorded. The excessive stress may be attributed to relic compression derived from past orogenies, though the role of continuing or modern earth movement of compressive type should not be overlooked; there is, indeed, much evidence of contemporary compression in the crust (Müller, 1964; Denham, Alexander and Worotnicki, 1979). For example, Isaacson (1957) reported that at a depth of 1056 m in one of the shafts of the Kolar Goldfield, southern India, the theoretical stresses ought to be 313.538 kg/cm^2 vertically and 134.98 kg/cm^2 horizontally, whereas, in reality, the measured stresses were 409.15 kg/cm^2 and 471.01 kg/cm^2 respectively. Expansion consequent on the release of inherent stress caused a shaft 3.81 m diameter at a depth of some 3048 m to decrease by 0.5334 cm in a north-south direction and by 1.16 cm east-west. Similar results have been obtained in the Snowy Mountains of New South Wales (Moyé, 1964).

Coates (1964) reported deformation in a tunnel some 90 m below the surface in southern Ontario. There was rapid expansion of the walls (contraction of the tunnel) in the first forty days after excavation followed by a period of several months of slower but similar changes so that, 240 days after exposure, there had been up to 4.6 cm of lateral expansion. Vertical movements were noted, but they were consistently small.

Many of the data and arguments outlined above as inconsistent with the offloading hypothesis can be used in support of the suggestion that sheet fractures and structures are associated with shortening. In addition, experimental work by Holzhausen (1989) has shown that lateral compression of partly confined blocks produces arcuate upwards strain trajectories (Fig. 2.14a). This finding

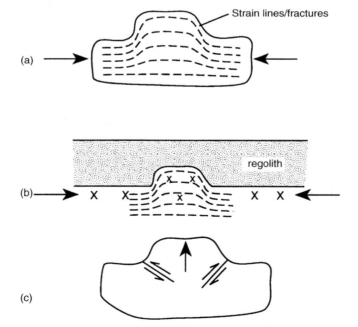

Figure 2.14. Result of compression of partly confined block: (a) laboratory situation (After Holzhausen, 1989), (b) result of compression of partly confined block in terms of the possible field reality, (c) suggested alternative result, possibly resulting from rapid or massive application of stress.

may be extrapolated and applied to bornhardts in terms of perhaps the most plausible general explanation of the form (Chapter 6). According to this two-stage concept, many bornhardts are initiated by differential weathering at the base of the regolith. At one stage they take the form of huge masses or compartments of intrinsically fresh rock projecting into the base of the regolith (Fig. 2.14b). Such projections are unconfined for they are surrounded and covered by regolith, which lacks strength. By analogy with Holzhausen's (1989) experimental work, compression would produce arcuate-upward stress trajectories, and eventually sheet fractures; though rapid application of stress could conceivably produce faulting and upthrust (Fig. 2.14c) rather than sheet fractures.

If sheet structure is associated with lateral compressive stress, if it is indeed arching, then the resultant domes are in some measure tectonic forms. As it is discussed later (Chapter 11), small-scale tectonism finds expression in several aspects of granite bornhardts and platforms, for example. In addition, some sheeting fractures are planes of dislocation (Fig. 2.10). In terms of compression, sheet fractures associated with surfaces recently sculptured by glaciers or by rivers (as depicted for example in Fig. 2.6) represent the release of stresses in parallelism with surfaces of least principal stress.

2.4 SUMMARY

Though any of several possible explanations of sheet jointing may be valid in particular areas, the hypothesis offering the best general explanation is that involving lateral compression, induced by horizontal stresses, either relic or modern, and the manifestation of stress patterns affecting brittle rock within, say, a kilometre of the surface when vertical loading is decreased by erosion. In these terms, sheet fractures and sheet structures are associated with tectonism. Such an explanation accounts for many details of the field evidence, is consistent with measured stress conditions, and offers a comprehensive view of the preservation of inselbergs and the sheet structure widely associated with them.

REFERENCES

Adams, F.D. 1954. The birth and development of the Geological Sciences. Dover Publications Inc. New York.
Brajnikov, B. 1953. Les pains-de-sucre du Brésil: sontils enracinés?. *Comptes Rendus Sommaire et Bulletin Societé Géologique de France* 6: 267–269.
Campbell, E.M. and Twidale, C.R. 1991. The evolution of bornhardts in silicic volcanic rocks, Gawler Ranges, South Australia. *Australian Journal of Earth Sciences* 38: 79–93.
Chapman, C.A. 1956. The control of jointing by topography. *Journal of Geology* 66: 552–558.
Choubert, B. 1974. Le Précambrien de Guyanes. Memoir Bureau de Recherches Géologiques et Minières 81.
Coates, D.F. 1964. Some cases of residual stress in engineering work. In Judd W.R. (ed.), The State of Stress in the Earth's Crust. Elsevier, New York. pp. 679–688.
Dale, T.N. 1923.The commercial granites of New England. United States Geological Survey Bulletin 738.
Denham, D., Alexander, L.T. and Worotnicki, G. 1979. Stresses in the Australian crust: evidence from earthquakes and in-situ stress measurements. Bureau of Mineral Resources *Journal of Australian Geology and Geophysic* 4, 289–295.
Fell, R., MacGregor, P. and Stapledon, D. 1992. Geotechnical Engineering of Embankment Dams. Balkema, Rotterdam.
Gilbert, G.K. 1904. Domes and dome structures in the High Sierra. *Geological Society of America Bulletin* 15: 29–36.
Holzhausen, G.R. 1989. Origin of sheet structure. 1. Morphology and boundary conditions. *Engineering Geology* 27: 225–278.
Isaacson, E. and de St.Q. 1957. Research into the rock burst problem on the Kolar Goldfield. *Mine and Quarry Engineering* 23: 520–526.
Lamego, A.R. 1938. Escarpas do Rio de Janeiro. Departamento Nacional da Produçao Mineral (Brasil) Serviço Geológico y Mineiro Boletim 93.
Merrill, G.P. 1897. Treatise on Rocks, Weathering and Soils. Macmillan, New York.

Moyé, D.G. 1964. Rock mechanics in the interpretation and construction of T.1 underground power station, Snowy Mountains, Australia. Annual General Meeting, Geological Society of America, Symposium on Engineering Geology, St Louis, 1958.

Müller, L. 1964. Application of rock mechanics in the design of rock slopes. In Judd W.R. (ed.), The State of Stress in the Earth's Crust. Elsevier, New York. pp. 575–598.

Ollier, C.D. and Pain, C.F. 1980. Actively rising surficial gneiss domes in Papua New Guinea. *Journal of the Geological Society of Australia* 27: 33–44.

Twidale, C.R. 1964. Contribution to the general theory of domed inselbergs. Conclusions derived from observations in South Australia. *Transactions and Papers of the Institute of British Geographers* 34: 94–113.

Twidale, C.R. 1973. On the origin of sheet jointing. *Rock Mechanics* 5: 163–187.

Twidale, C.R., Vidal Romaní, J.R., Campbell, E.M. and Centeno, J.D. 1996. Sheet fractures: response to erosional offloading or to tectonic stress?. *Zeitschrift für Geomorphologie*, Supplement Band 106:1–24.

Vidal Romaní, J.R., Twidale, C.R., Campbell, E.M. and Centeno, J.D. 1995. Pruebas morfologicas y estructurales sobre el origen de las fracturas de descamacion. *Cadernos do Laboratorio Xeolóxico de Laxe* 20: 307–346.

White, W.A. 1945. Origin of granite domes in the south-eastern Piedmont. *Journal of Geology* 53: 276–282.

3

Weathering

3.1 DEFINITION AND SIGNIFICANCE

Weathering may be defined as the disintegration or decay of rocks *in situ* and in the range of ambient temperatures found at and near the Earth's surface (Winkler, 1965). Some weathering processes are physical or mechanical and result in the break down or fragmentation of the rock. Others are chemical and involve the alteration of one or more of the constituent minerals. Biota also make important contributions to both types of weathering, and indeed, several processes, of various kinds, commonly work together to produce a weathered mantle or regolith, which with the addition of organic materials becomes a soil. The lower limit of significant or detectable weathering is called the weathering front.

Weathering is an essential precursor to erosion: without preliminary weathering of the rock, there would be little erosion. Some weathering processes, however, result not in the weakening of the rock but rather in its cementation and induration, through the development of concentrations of minerals called duricrusts of which laterite, ferricrete, bauxite, silcrete, calcrete and gypcrete are well-known and widely developed examples. None, however, is peculiar to granitic terrains. On the other hand, many of the minerals forming the duricrusts are allochthonous, being transported in groundwaters, in rivers, or on the wind, so that, combined with locally derived contributions, laterite and silcrete, for example, are well represented in granite landscapes. Where the duricrusted surface is dissected, plateau forms are characteristic (Fig. 1.1h). Where intact, the duricrust forms a protective carapace, as for example on northern Eyre Peninsula, South Australia, where the rolling granite plain carries a veneer of calcrete which has not only stabilised the surface but has induced a weak karst.

3.2 PHYSICAL DISINTEGRATION

Earlier investigators set great store by physical processes, and temporal variations in insolation, involving alternations of heating and cooling (Griggs, 1936), for example, were widely held responsible for granular disintegration and for flaking and spalling (widely referred to as exfoliation) in granite. The argument is plausible. Granite consists of minerals of different colours with different coefficients of heating and therefore of expansion. The stresses generated by alternations of heating and cooling were considered sufficient to cause fragmentation. In addition, rocks are poor conductors of heat, so that rocks exposed to the Sun would be heated and would expand, whereas deeper sectors would not, and it was considered that as a result the outer skin would separate from the inner, resulting in flaking and spalling. It has long been realised that granite expands on heating. Indeed, this knowledge was applied in quarrying in ancient Egypt and in India. Also, the intense, though ephemeral, heat of forest or bush fires unquestionably causes flaking of exposed surfaces (Fig. 3.1); and the heat generated in nuclear explosions has similar effects.

50 *Landforms and Geology of Granite Terrains*

Figure 3.1. Spalling of granite due to bush fires (a) near the Portugal-Galicia frontier line of Portela d'Home; (b) at the Devil's Marbles, Northern Territory. Most fire-induced heating produces thin flakes and slivers, only 2–3 mm thick, but here the plates spalled off are several centimetres in thickness.

But field observations suggest that, at the very least, moisture-related weathering acts far more rapidly and widely than do insolational effects, and many of the features attributed to heating and cooling must be due to other processes. Flaking and spalling demonstrably extend to depths of many metres, far beyond the effect of diurnal or even secular temperature changes (Fig. 3.2). Again, in the Nile valley and in the adjacent deserts the surfaces of granite blocks exposed to the Sun's rays (Barton, 1916) for a few thousands of years show no signs of alteration or disintegration,

Figure 3.2. (a) Flaking of granite around corestones at Palmer, eastern Mt Lofty Ranges, South Australia. (b) Spalling in granodiorite in the Snowy Mountains, New South Wales.

whereas their shady undersides, and especially any parts covered by desert sand or river silt – both of which hold moisture, in the case of the desert sand most commonly derived from dew or fog – show clear signs of alteration. Such observations do not preclude disintegration due to insolation, but they strongly suggest that chemical changes due to contact with water act more rapidly. Also, in Galicia, there are rock carvings 5,000–7,000 years old. Some of them were buried by soil. Others were not. Those which were covered are more decayed than those that remained exposed, because the soil held moisture which attacked the rock surface.

On the other hand, temperature oscillations in the presence of moisture apparently cause fragmention, both above freezing, and around freezing point (Grawe, 1916). Well-fractured rocks are

Figure 3.2. (c) Spalling in granodiorite in the Snowy Mountains, New South Wales.

Figure 3.3. (a) Scree slopes of granite blocks and fragments, Panticosa, Pyrenees, northeastern Spain.

especially susceptible, and the clitter and screes, consisting of angular blocks and plates (Fig. 3.3), found in granitic terrains in cold regions are widely attributed to freeze-thaw alternations. Flaggy granite (pseudobedding) is especially well-developed in cold regions and may also be an expression of freeze-thaw activity, though the water and ice may exploit strain zones in the country rock. Various aspects of the freeze-thaw mechanism have been questioned from time to time (White, 1973) (for instance, pressures in adjacent partings are opposed, and only oscillations around freezing point are effective) but the field evidence pointing to the reality of frost shattering is compelling.

Crystallisation of salts such as halite and gypsum (Bradley, Hutton and Twidale, 1972) has been shown experimentally to exert enough force to disrupt even fresh granite. The mechanism, known

Weathering 53

Figure 3.3. (b) These plates of granite have been produced by frost riving in the Rocky Mountains of Colorado.

Figure 3.4. Disruption of granite slabs by tree roots, Wudinna Hill, northwestern Eyre Peninsula, South Australia.

as haloclasty, has been invoked in explanation of tafoni and alveoles (see Chapter 10), as well as general disintegration, in arid and semi-arid regions both hot and cold, coastal and interior (Joly, 1901).

Tree roots have obviously disturbed and exposed granite slabs and blocks (Fig. 3.4) and at a smaller scale the roots or hyphae of lichens disrupt crystals and fragments (Fry, 1926; McCarroll and Viles, 1995). Algae and bacteria also bore into crystals thus creating pathways along which water can penetrate. Indeed, there is increasing evidence that nannobacteria or midget or dwarf bacteria

(Folk 1993, 1994; Folk, Noble, Gelato and McClean, 1995) are exceedingly important in the weathering of all rocks, including granite (Pedersen, 1997).

Erosional offloading is widely cited as another cause of physical breakdown, resulting in sheet fractures and sheet structure, but the arguments against it are several, and have been outlined in Chapter 2.

3.3 CHEMICAL ALTERATION

There is some suggestion that chemical reactions take place at grain boundaries in dry conditions, but infiltration by moisture and gases produces pronounced and widespread alteration by such processes as oxidation, reduction, carbonation, solution, hydration and hydrolysis. Because of its molecular structure, water is an ideal solvent. No other liquid can dissolve such a variety and volume of solutes. It has been claimed that solution is essential to chemical weathering, not only because of its widespread direct effects, for all minerals are soluble to some extent, but also because it prepares crystal structures for further reactions. Some workers emphasise solution, others consider hydrolysis to be the most effective of these water-related processes, and yet others favour hydration. Suffice is to say that all play their part and that all may be especially important in particular circumstances. Hydration implies dissociation of water and the release of hydrogen ions. Because of their high energy and small ionic radius they are active in substitution and they readily enter and disrupt crystal lattices. Plant roots assist by concentrating hydrogen ions. On the other hand, alkaline conditions are prevalent in arid and semi-arid regions like the Australian interior, and silicate minerals and quartz, which are important constituents of granitic rocks, are, as Joly (1901) showed a century or so ago, more reactive to alkaline than to neutral solutions.

In general terms, a granitic rock consisting of quartz, a potassium feldspar, some plagioclase and a mica is first broken down to a granite sand or grus (or fine gravel – see Chapter 5) consisting of fragments of quartz and feldspar. Some workers differentiate between mechanically disintegrated grus and chemically altered growan, but there is a gradation between granite sand containing only slightly altered minerals, and the gritty clay that is the common end-product of the weathering of granite; it is difficult to distinguish between the two types of weathering product in the field. The quartz persists for a long time, suffering only slow dissolution, but eventually it is too dissolved because at some sites quartz fragments present in some horizons are absent from those above. Also, as is described in Chapter 10, siliceous speleothems are deposited in open sheeting fractures and other apertures, demonstrating that silica must go into solution before being reprecipitated. Much, perhaps most, of the silica is derived from the breakdown of silicates such as the feldspars, but siliceous speleothems are also found in openings in sandstone in many parts of the world and notably in the Roraima Plateau of Venezuela, showing that quartz goes into solution.

Water reacts with the mica and feldspars to produce clays. The character of the clay varies with local and regional circumstances (such as whether the system is open or closed, whether the profile is well drained or not) so that if potash resulting from the hydration of orthoclase or microcline remains in the system, illite is produced, but if it is evacuated kaolinite is formed (McFarlane and Heydemann, 1984). Most commonly, the weathering of granite through contact with water results in the formation of kaolinite.

3.4 THE COURSE OF WEATHERING IN GRANITE

Near the land surface in contact with soil, the differential weathering of the various constituents of granite, and in particular the preferential weathering of feldspar and mica, leaves crystals of quartz, and phenocrysts of orthoclase and microcline, in microrelief, producing a pitted surface (Fig. 3.5). Such surfaces denote recent exposure, though how recent is not known and it may well vary from one environment to another. The extent, and particularly the depth, of such recent exposures as indicated by pitting, is in some places surprising. For example, the sidewalls of a valley

Figure 3.5. Pitted surfaces in granite: (a) Sketch showing the development of pitting, (b) detail of pitted boulder surface at Mt Bundey, near Darwin, Northern Territory, (c) on boulder near Tampin in West Malaysia.

Figure 3.5. (d) Showing overall effect where exposed (the darker zone) by the removal of soil from the base of the rock, at Murphys Haystacks, northwestern Eyre Peninsula, South Australia. (e) Valley on Domboshawa, near Harare, Zimbabwe, showing pitting extending some 2 m up valley sidewalls.

on Domboshawa, a granitic bornhardt near Harare in central Zimbabwe (Fig. 3.5e), are pitted to a depth of about two metres. Some areas of pitted surface have clearly been eliminated by recent wash, but a detrital fill of significant thickness was formerly present.

Elsewhere, the course of weathering can be deduced from an examination of weathered mantles or regoliths. Regoliths develop from the surface downwards. The weathering front, the interface between the regolith and the intrinsically fresh rock, may be diffuse, but in granite is commonly sharp and takes the form of either a distinct plane or a narrow zone of transition. This abrupt change reflects the crystallinity of the rock and also the susceptibility of feldspar and mica to water attack. The fresh rock is impermeable, but once water has penetrated along crystal cleavages, or microfissures, alteration takes place within the body of the rock. The permeability of the

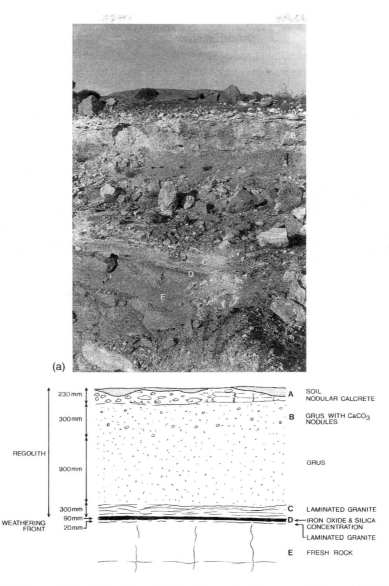

Figure 3.6. (a) Photograph (b) sketch of 2 m thick regolith exposed in quarry at Yarwondutta Rock, near Minnipa, northwestern Eyre Peninsula, South Australia. A – soil; B – Grus with limestone nodules; C – laminated granite; D – FeO-rich zone; E – Fresh granite.

rock increases dramatically, so that more water penetrates and more alteration rapidly occurs. The weathering front descends from the surface to depths of a few metres (Fig. 3.6), or, in humid tropical lands, several scores, or even a few hundreds, of metres (Branner, 1896). Thus, in many places the continents carry a skin of regolithic material which is important not only by virtue of the surface soil, but also because it holds water. Reactions between regolithic moisture and the underlying bedrock are responsible for many familiar landforms both in granitic terrains and elsewhere.

Because the regolith develops from above and penetrates down (and laterally) into the country rock, the initial phase or stage of weathering is found at the weathering front and successively higher zones indicating stages of more advanced weathering. At many sites, both in granite and in other rock types, the rock immediately above the weathering front (in the corestones that become boulders – see Chapter 5 – adjacent to the kernels of fresh rock) is laminated, flaked or spalled. Such changes have been attributed to moisture penetrating along crystal cleavages and microfissures, and causing hydration or hydrolysis (Fig. 3.2). Though the changes effected are acknowledged to be minute, they have been assumed to be sufficient to produce volume increase and physical disruption. Flaking (or micro-sheeting) due to the hydration of biotite, for example, causes it to expand. Where the crystal is confined or buttressed (Folk and Patton, 1982), this expansion is converted to arching which finds expression in flaking, scaling or lamination, and eventually, given the overall expansion caused by hydration and related reactions in grus the desintegrated and altered rock.

In addition, however, once laminae have been formed, water can readily penetrate into the rock, and not only continue the physical breakdown of the rock – an example of a positive feedback or reinforcement mechanism – but also effect various chemical changes. The flakes are fragmented and the mica and feldspar are gradually altered to clay, the character of which depends on conditions within the regolith, but which is commonly kaolinite. Eventually, the quartz must be also dissolved. Salts released by weathering are illuviated and, translocated by descending meteoric waters and shallow groundwaters, tend to accumulate at the base of the regolith, just above the impermeable fresh rock. In particular iron oxides and amorphous silica derived from the alteration of micas, feldspars and quartz (Siever, 1962) are concentrated at the weathering front. In the Yarwondutta Quarry, Western Australia (Fig. 3.6) iron and silica (which have replaced plant roots) accumulated at the weathering front are 2–3 times as abundant as in the fresh rock.

In effect, water rots granite, changing it from a cohesive tough rock to a weak, puggy, gritty clay. Well might MacCulloch (1814, p. 72) refer to the alteration of granite as a "gangrenous process".

3.5 CONTROLS OF WEATHERING

Several factors influence the type and rate of rock weathering. All else being equal, the susceptibility of minerals to alteration is the same as the order in which they crystallise out from an igneous melt. The higher temperature minerals are in greater disequilibrium with the Earth's surface environment than those that crystallise out at lower temperatures. Thus, the composition of the rock (Bowen, 1918) strongly influences its rate of alteration, with rocks rich in such minerals as olivine, augite and hornblende more susceptible than those composed of quartz and potash feldspars, for example: all else being equal, basalt, norite and gabbro are more readily weathered than dacite and granite (Hutton, Lindsay and Twidale, 1977). Granite is compositionally a resistant rock; yet different granites vary in their weathering characteristics because of variations in composition. For instance, biotite is a weak link, and biotite granites in French Guyana for example, are preferentially weathered. The granites of the Karkonosze Mountains of southern Poland are resistant because they lack biotite and other ferromagnesian minerals. Even within massifs compositional variations find morphological expression, the vertical western face of the Pão de Açucar, in Rio de Janeiro, southeastern Brazil, for example, being partly due to the exploitation of a biotite-rich zone. In parts of Corsica, bluffs and other steep slopes eroded in rocks rich in ferromagnesian minerals stand in contrast with the more gentle inclines of granodiorite and monzonite terrains. On Haytor, on eastern Dartmoor, southwest England, the Giant Granite appears to be more resistant than the Blue, with the result that a shallow alcove is associated with the latter (Fig. 3.7); though it could reflect more concentrated attack by soil moisture at a time when the regolith stood higher on the flanks of the residual.

In a more general sense, granodiorite is not only by far the most common of the granitic rocks, but consisting, as it does, of over 40% feldspar, predominantly plagioclase, which is readily susceptible to reactions with water and resultant alteration to clays, it is, of all those in the granitic domain, the rock most vulnerable to weathering by reason of its composition.

But composition is in many places overridden by fracture density. Fractures are avenues of water penetration and thus of weathering. Variations in fracture density and the attitude of partings

Weathering   59

Figure 3.7. On Haytor, eastern Dartmoor, southwestern England, the Giant Granite forms most of the hill, but Blue Granite is exposed near the base of the slope. It is evidently more susceptible to weathering than the overlying rock for a shallow alcove has been formed.

Figure 3.8. Contrasted textures of topography due to different fracture densities in granitic rock, Fafião Valley, southern Galicia.

frequently find expression in the landscape (Fig. 3.8). Massive compartments of rock, lacking, or with few open, fractures, withstand weathering much more readily than those riddled with open partings. Thus, in the foothills of the western Sierra Nevada, near Fresno, in California (Larsen, 1948), there are hills underlain by norite, which on account of its mineralogy ought readily to be weathered, but fracture density is low, water cannot penetrate the rock mass and it remains upstanding. Similarly, basalt ought to be altered rapidly and intensely on contact with water. The extensive and widely distributed basalt plateaux, which are prominent in many parts of the world, owe their survival to their perviousness arising from well-developed systems of columnar joints and flow partings.

Quartz grains in granite are commonly riddled with cracks or microfractures, due to tectonic stresses. Crystals outer parts are in disequilibrium and their lattices are more readily penetrated by other atoms and molecules so that zones of strain are in many places preferentially weathered and eroded.

Rock texture also finds expression in the landscape. Assuming water penetration, fine-grained rocks ought to be more susceptible to alteration than coarse-grained, because there is a larger area of crystal surface per unit volume, and there are greater areas of contact between minerals where reactions can take place. It has been claimed that fine-grained granites in southern China and southern Poland are more resistant than adjacent coarse-grained facies. But in Brazil it is reported that in some areas coarse granite underlies hills, though elsewhere the coarse rock is more readily weathered than an adjacent fine-grained rock.

Other factors such as fracture density, intervene, and moreover all is relative: south of Karibib, in central Namibia, a pegmatite vein has been more rapidly weathered than the granite into which it is intruded (Fig. 3.9a), but on adjacent slopes eroded in biotite schist the same vein forms a low rib. On the other hand, an aplitic sill intruded into granite at Paarlberg, near Cape Town, South Africa, is evidently more resistant than the presumably metamorphosed rock to either side, but is not as tough as the fresh granite (Fig. 3.9b). And an aplitic sill on Freeman Hill, northwestern Eyre Peninsula, South Australia, is in places clearly more resistant than the host rock, for it forms a low wall (Fig. 3.9c), though nearby (Fig 3.9d) it stands flush with the granite surface.

Climate is an important determinant of the type and rate of weathering. Because water is involved in many weathering processes, because most chemical reactions take place more rapidly at high rather than low temperatures, and because organic acids and biota are more abundant in warmer environments, the humid tropics provide optimal conditions for alteration. Such conditions were more widespread in the geological past than at present. Not only are extensive thick regoliths found in the humid tropics but there is clear evidence that in such areas many minerals, including some granite-forming minerals, are detectably altered within a few years of exposure to the elements. Thus, micas show signs of decay within a few decades of exposure in places like Madagascar and southeastern Brazil, and feldspars (admittedly plagioclase) in a few centuries in Indonesia and the Antilles (Goldich, 1938; Loughnan, 1969; Ollier, 1969). On the other hand, frost

Figure 3.9. (a) This pegmatite vein exposed on a granite slope near Karibib, central Namibia, has been weathered more rapidly than the host rock.

Figure 3.9. (b) Aplitic sill intruding granite at Paarlberg, near Cape Town, South Africa. Note that the sill is more resistant than the presumably metamorphosed rock immediately adjacent to it, but less resistant than the fresh granite. (c) Aplitic sill level with granite surface, Freeman Hill, northwestern Eyre Peninsula. (d) This aplite vein, also at Freeman Hill, is more resistant than the granite into which it is intruded and stands up as a low wall (except where, as in the foreground, the aplitic blocks have been removed for use in local buildings).

shattering acts rapidly and is very effective in cool latitudes and altitudes, and biota are destroying carbonate and other rocks at measurable rates in coastal zones in the Mediterranean and in the tropics.

Time is also germane to the problem of weathering, for given long exposure, even slow-acting processes and reactions assume importance. As many granites are emplaced in the ancient shield lands or in old orogens, this is a factor of some significance in the granite context. Moreover, as tectonic, climatic and topographic conditions have changed, so may biota and groundwater circulation, for example, have changed in time.

Thus, several factors impact on the rate and type of weathering active in given areas at particular times. Some workers (Birkeland, 1974) emphasise rock composition, others climate, but because of the importance of water, and the chemicals and biota it carries, in rock weathering, many point to fracture density as the most important single control of weathering. The salient point, however, is that most granites are inherently susceptible to moisture attack and are readily weathered and eroded.

Weathering processes in the granite context are further discussed in relation to various specific landforms in the following chapters.

REFERENCES

Barton, D.C. 1916. Notes on the disintegration of granite in Egypt. *Journal of Geology* 24: 382–393.
Birkeland, P.W. 1974. Pedology, Weathering and Geomorphological Research. Oxford University Press, London.
Bradley, W.C., Hutton, J.T. and Twidale, C.R. 1972. Role of salts in the development of granitic tafoni, South Australia. *Journal of Geology* 86: 647–654.
Bowen, N.L. 1928. The Evolution of the Igneous Rocks. Princeton University Press, Princeton, New Jersey.
Branner, J.C. 1896. Decomposition of rocks in Brazil. *Geological Society of America Bulletin* 7: 255–314.
Folk, R.L. 1993. Interaction between bacteria, nannobacteria, and mineral precipitation in hot springs of central Italy. *Géographie Physique et Quaternaire* 48: 233–246.
Folk, R.L. 1994. SEM imaging of bacteria and nannobacteria in carbonate sediments and rocks. *Journal of Sedimentary Petrology* 63: 990–999.
Folk, R.L., Noble, P.J., Gelato, G. and McClean, R.J.C. 1995. Precipitation of opal-CT lepispheres, chalcedony and chert nodules by nannobacteria (dwarf bacteria). Geological Society of America, Annual Meeting, New Orleans. Abstracts A-305.
Folk, R.L. and Patton, E.B. 1982. Buttressed expansion of granite and development of grus in central Texas. *Zeitschrift für Geomorphologie* 26: 17–32.
Fry, E.J. 1926. The mechanical action of corticolous lichens. *Annals of Botany* 40: 397–417.
Goldich, S.A. 1938. A study in rock-weathering. *Journal of Geology* 46: 17–58.
Grawe, O.R. 1936. Ice as an agent in rock weathering: a discussion. *Journal of Geology* 44: 173–182.
Griggs, D.T. 1936. The factor of fatigue in rock exfoliation. *Journal of Geology* 44: 783–796.
Hutton, J.T., Lindsay, D.S. and Twidale, C.R. 1977. The weathering of norite at Black Hill, South Australia. *Journal of the Geological Society of South Australia* 24: 37–50.
Joly, J. 1901. Expériences sur la denudation par dissolution dans l'eau douce et l'eau de mer. *Comptes Rendus Congrès Géologique International* (Paris) 8 (2): 774–784.
Larsen, E.S. 1948. Batholith and associated rocks of Corona, Elsinore and San Luis Rey quadrangles, southern California. Geological Society of America Memoir 29.
Loughnan, F.C. 1969. Chemical Weathering of the Silicate Minerals. Elsevier, Amsterdam.
MacCulloch, J. 1814. On the granite tors of Cornwall. *Transactions of the Geological Society* 2: 66–78.
McCarroll, D. and Viles, H. 1995. Rock-weathering by the lichen Lecidia auriculata in an Arctic Alpine environment. *Earth Surface Processes and Landforms* 20: 199–206.
McFarlane, M.J. and Heydemann, M.T. 1984. Some aspects of kaolinite dissolution by a laterite-indigenous micro-organism. *Geo-Eco-Trop* 8: 73–91.
Ollier, C.D. 1969. Weathering. Oliver and Boyd, Edinburgh.
Pedersen, K. 1997. Microbial life in deep granitic rock. *Episodes* 20 (1): 7–9.
Siever, R. 1962. Silica solubility 0–200°C, and the diagenesis of siliceous sediments. *Journal of Geology* 70: 127–150.
White, S.E. 1973. Is frost action really only hydration shattering? A review. *Arctic and Alpine Research* 8: 1–6.
Winkler, E.M. 1965. Stone: Properties, Durability in Man's Environment. Springer, New York.

4

Plains – the expected granite form

4.1 WEATHERING AND SURFACES OF LOW RELIEF

Granite in contact with water is readily and rapidly weathered. Two of the common rock-forming minerals found in granite, mica and feldspar, are readily, and, in geological terms, rapidly, attacked. Shallow groundwaters are ubiquitous. Granites are typically well fractured, for in addition to orthogonal systems and sheet fractures, there are many random partings and numerous microfissures. Granites include minerals with a well-developed cleavage. Many granite gneisses are foliated and have mineral banding. Thus, there are various and varied avenues permitting the entry of water, and with it, chemicals and biota. Granites are major components of the ancient shield lands which are the nuclei of the continents. Such masses have been exposed to meteoric waters, and particularly groundwaters, for hundreds of millions of years. For these reasons they have been weathered and worn down to regions of subdued relief. Also, physical contrasts are introduced through the development of a regolith, which is generally more susceptible to erosion than on fresh granite; and many plains have been formed beneath and exposed as a result of the stripping of the regolith.

Plains are thus the expected and areally the most extensive granite landform. Despite the understandable interest evidenced in the positive relief features developed on granite, plains are by far the most characteristic landform developed on granitic bedrock. Plains are indeed an essential component of such well-known, and typically granitic features as Inselberglandschaften, for it is the contrast between the virtually featureless sweeping plains and the steep-sided residuals that endows the landscapes with their dramatic aspect (Fig. 1.2e). The plains of inselberg landscapes vary both morphologically and genetically. An early student of such features, Passarge (1904), classified the plains of inselberg landscapes according to their origin and age, and named them after typical African occurrences. The Banda type is, according to Passarge (1904), scoured by the wind and consists of an assemblage of bedrock plains and depressions brought into conformity with degradational elements by aeolian deposits. The Rovuma type is similar but was shaped during desert conditions during the later Mesozoic, and was subsequently buried and exhumed. The Kordofan type includes some aeolian elements but is shaped mostly by rivers, and the plains of Adamaua type are shaped entirely by rivers and streams.

It is rare to have both the evidence and the confidence to interpret in such detail, and it is more practical to classify plains according to their morphology and stratigraphic history.

4.2 PLAINS OF EPIGENE (SUBAERIAL) ORIGIN

Plains eroded in granite are well represented in tropical and subtropical lands, though they are by no means restricted to them, and they are typical of the shield lands. Plains developed by epigene or subaerial processes on granitic rocks vary in both extent and morphology. Pediments are a feature of the piedmont and are of limited areal extent, whereas planation surfaces, in places rolling or undulating, elsewhere flat and featureless, occupy huge areas (Mabbutt, 1966, 1978).

4.2.1 *Rolling or undulating plains*

A peneplain is a rolling or undulating erosional surface of low relief and of regional extent formed by weathering, wash and river work. The broad interfluves are gently convex-upward in form. According to Davis (1899, 1909), slopes are lowered once streams have achieved their initial major incision and when maturity has been reached and passed. Divides are lowered more rapidly than the streams are incised, and a surface of low relief is developed.

Though principally of erosional origin, flood plains due to deposition extend along the main drainage lines. It is difficult to demonstrate the slope decline implied in the Davisian model though several workers have produced statistical evidence which is consistent with the scheme.

(a)

(b)

Figure 4.1. Rolling plains in granite (a) southwest of Western Australia. (b) Saldanha, northwestern Cape Province, South Africa.

Davis (1899, 1909) himself regarded peneplains as products of what he called normal erosion, that is, river work in temperate humid conditions. Davis (1909) cited parts of the western Great Plains in Montana, and of Siberia as examples, but most of the peneplains he and later workers recognised were, in fact, palaeoplains preserved on upland crests.

Planation surfaces morphologically similar to the peneplain adduced in the theoretical model are developed on granitic rocks in the southwest of Western Australia (Fig. 4.1a) and on central and northern Eyre Peninsula, South Australia. Similar features are found also in southern Africa, for instance, in Northern Province, in Western Cape Province, in parts of central Namibia, and in central Brazil.

Such granitic peneplains are underlain partly by fresh rock, though mostly by weathered material. Some, however, differ in detail from the model deduced by Davis (1899, 1909). In many of the areas cited rivers are intermittent or seasonal in their flow, and are braided in pattern. The peneplains of northern and central Eyre Peninsula are developed in weathered, permeable granites (Bourne, Twidale and Smith, 1974). They also carry a discontinuous but extensive carapace of calcrete developed during the later Pleistocene (most of the constituent lime was carried on the wind from the extensive dune calcarenites of the west coast). Thus, rainwater readily permeates into the subsurface. For this reason, there are few surface streams, save during and immediately following rains. But in general terms the examples cited conform to Davis' model. Again, the residuals that stand above peneplains are not everywhere Davisian monadnocks that rise gently from the surrounding planate surfaces; on the contrary, in many places inselbergs, pediments and peneplains coexist (Fig. 4.2).

Peneplains are not restricted to temperate lands. Indeed, they are not confined to any conventionally defined climatic region. Like pediments, they are characteristically developed on weak rocks, typically argillaceous sediments or weathered crystalline rocks. It is difficult to determine whether slope decline or backwearing has been involved in their development. On general grounds it can be argued that in areas of weak rocks downwearing ought to have been dominant, but there may have been local or ephemeral complications, such as aridity and the development of a gibber veneer, or the development of a duricrust, in either case implying a caprock and scarp retreat. The end result was the lowering of the land surface, but there is no means of demonstrating past slope behaviour in the field.

Surface waters infiltrate into the regolith where they not only continue the processes of weathering but also transport some of the products of alteration (Davis, 1963). After heavy rains, fines may be flushed through the system, but more significantly, salts are taken out in solution. At present, and on average, of the order of 4 billion metric tonnes of carbonates, silica and sulphates, etc.

Figure 4.2. (a) Ucontitchie Hill and surrounding mantled pediment, Eyre Peninsula, South Australia.

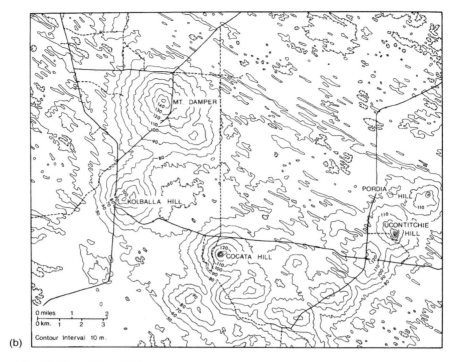

Figure 4.2. (b) Map of Ucontitchie area.

are transported annually in streams alone (Livingstone, 1963). This figure takes no account of the immense volumes of salts in solution that are transported to the oceans in groundwaters nor those that are carried in internal drainage systems. The volume of such dissolved solids varies greatly from region to region, from some 15.5 metric tonnes per square kilometre in Australia to 47 in Europe. What is clear is that vast quantities of dissolved minerals are carried in rivers and groundwaters, and that such subsurface erosion could materially contribute to surface lowering following volume decrease and compaction in the regolith.

4.2.2 *Pediments*

By definition, pediments are gently inclined, cut bedrock surfaces located in the piedmont zone (Figs 4.2 and 4.3). Though many pediments carry a regolith veneer, the surface form reflects the slope of the bedrock surface. This is the essential difference between pediments and alluvial fans, for the form of the latter is a consequence of deposition, and the inclination of the surface is a function of the gradient of the streams responsible for the transportation and deposition of the debris. On granitic rocks the slope of pediments varies between 0.5° and 7°, but they are typically inclined at 0.5°–2.5° with respect to the horizontal. Most are gently concave upward, many are rectilinear, and a few are convex. But whatever their geometry, pediments meet the backing escarpment in an abrupt break of slope called the piedmont angle or pediment nick. Three types of pediment have been recognised: those with a cover of allochthonous debris, those with a mantle derived primarily from the weathering of bedrock *in situ*, and rock pediments or platforms which are essentially devoid of a cover of unconsolidated material. Of these, the first type is essentially restricted to sedimentary terrains. Such covered pediments are rarely formed in granite. Some are developed, as near Usakos, in central Namibia (Fig. 4.4), but they are unusual. On the other hand, mantled and rock pediments (the latter also known as platforms – see Chapter 8) are well-developed on granitic rocks.

Figure 4.3. (a) Pediment profile in gneissic rocks, central Namibia. (b) Small sand mantled granite-gneiss pediments in northern Namaqualand (Western Cape Province).

Figure 4.4. Pediment in granite, but with cover of travertine-cemented river conglomerate, near Usakos, central Namibia.

- The basic unit of the mantled pediment is the low-angle, fan-shaped, cone segment cut in bedrock, but carrying a thin veneer of sand, partly originating in the backing upland and transported either by small streams (where the cone occurs at the mouth of a valley) or by wash (where it is located at the base of a bluff), but mostly derived from the weathering of the bedrock of the piedmont. Some such part-cones stand in isolation, others have merged with similar features to form either low-angle cones that surround isolated residuals or aprons such as those that front inselberg ranges in many parts of Australia, southern California, and Namibia and Namaqualand in southern Africa (Figs 4.2, 4.3 and 4.5). Some such coalesced pediments are called pan fans, or, perhaps more commonly, pediment aprons.

Figure 4.5. (a) Mantled pediment in the southern piedmont of Granite Mountains, southern California. The mantle has been partly stripped, exposing closely packed corestones. (b) Corestones in regolith exposed in excavation in pediment near Granite Mountains, southern California.

Such mantled pediments are cut in granite, but the intrinsically fresh bedrock is masked or mantled by a veneer that consists largely of weathered material *in situ*. In places, as, for example, bordering the channels of ephemeral streams, the bedrock is masked by a thin layer of stratified alluvium, but the mantle consists overwhelmingly of grus. Such pediments have been described from areas of pronounced winter rains, such as the Mediterranean region and Eyre Peninsula, South Australia; in areas of continental and monsoonal climate such as Korea and in Japan, and in areas that experience only episodic rains – desert and semi-desert regions like the Sahara, the American West and Southwest, central Australia, and so on.

– Rock pediments or platforms are planate rock surfaces. McGee (1897) recorded his astonishment on encountering these forms in Arizona late last century. They were different from anything previously recorded:

"At first sight the Sonoran district appears to be one of half-buried mountains, with broad alluvial plains rising far up their flanks, and so strong is this impression on one fresh from humid lands that he finds it difficult to trust his senses when he perceives that much of the valley-plain area is not alluvium but planed rock similar or identical with that constituting the mountains ...

During the first expedition ... it was noted with surprise that the horse-shoes beat on planed granite or schist or other hard rocks in traversing plains 3 or 5 miles from mountains rising sharply from the same plains without intervening foothills ... " (McGee, 1897, pp. 90–91).

Rock pediments in granite are gently inclined and are typically dimpled and grooved. Many carry remnants of a regolith (Fig. 4.6). Some fringe uplands, others stand in isolation as the just-exposed crests of emerging domes or as the reduced remnants of once high masses (Fig. 4.7), though some would not classify the latter as pediments because they lack a backing scarp and hence a piedmont angle.

– How pediments form or may have formed has given rise to considerable debate. For many writers, pediments developed on crystalline rocks are, in broad view, a consequence of scarp recession. Thus, Howard (1942, p. 134), who worked in crystalline terrains in the American Southwest, wrote that "... *the development of the pediment depends on the recession of the base of the slope* ...", and Pugh (1956, p. 28), who investigated granite landforms in Nigeria, concluded that "... *a mountain mass with a well-developed upper surface will shrink slowly by scarp retreat, with the development of bounding pediments.*"

But this view is open to serious questioning for, regardless of the precise nature of the formative processes, there is compelling evidence to suggest that though in many cases backwearing

Figure 4.6. Granite platform with patches of a thin regolith, between Corrobinnie Hill and Peella Rock, northern Eyre Peninsula, South Australia.

70 *Landforms and Geology of Granite Terrains*

Figure 4.7. Isolated granite platforms, (a) the exposed crests of domes in western Pilbara, Western Australia. (b) In the piedmont of Ampidianambilahy, Andringitra Massif, Madagascar.

related to scarp-foot weathering has taken place, it amounts at most to only a few tens of metres (see Chapter 6) whereas pediments may be scores, or even a few hundreds of metres across. The residual boulders standing on some pediments are flared, suggesting that they were weathered during a period of topographic stability when the weathering front was lowered (Chapter 8). Then followed a phase of erosion, during which the friable grus was evacuated to expose the boulders shaped by subsurface moisture attack. Thus, on the Waulkinna pediment, and at Houlderoo, both granitic and both located in the southern piedmont of the Gawler Ranges, South Australia, the evidence suggests that there has been a lowering of the surface by some 2 m (Fig. 4.8).

The concentration of moisture in the scarp-foot zone has led to the development of moats of especially deep weathering – incipient scarp-foot depressions – as well as flared forms and cliff foot caves or tafoni, all of which argue relative stability of marginal slopes and weathering zones. It is true that, in appropriate settings, wind-driven waves and associated marine agencies, frost action and glaciers have caused local wearing back of basal slopes and escarpments, but such activities are specific to certain environments, with which pediments are not commonly associated.

Figure 4.8. Conical residual or coolie hat, 2 m high and weathered in granite, on Houlderoo pediment (part mantled, part rock), southern piedmont of the Gawler Ranges, South Australia.

In addition to this evidence suggestive of the stability of bounding scarps, many of the inselbergs surmounting or backing the pediments are of limited areal extent (Fig. 4.2). Some are massive, others are well-fractured and pervious, yet all are fringed by pediments, despite their generating little overland flow. The regolith consists largely of grus *in situ*. For this reason any suggestion that the pediments adjacent to granitic inselbergs are comparable to those described from some sedimentary terrains and due to lateral corrasion by divaricating streams must be rejected.

The field evidence suggests that mantled and rock pediments are end members of a continuum (Bryan, 1922, 1936). The mantled forms carry a veneer of weathered rock smoothed by wash and rill work. The rock pediments are exposed areas of the weathering front and are due to mantle-controlled planation. They are mantled pediments from which the regolith has been stripped. Evidence for this conclusion derives from the nature of the mantle exposed on and adjacent to the platforms; from the corestones still *in situ* and embedded in the grus at some sites and scattered over the pediments as boulders in others; from the physical continuity of the platforms and such features as flared slopes that are demonstrably a particular form of weathering front; and from the pitted, grooved and dimpled morphology of the platforms, these minor features being initiated at the weathering front beneath the regolith (Chapters 8 and 9). The relative rate of lowering of the weathering front and the mantle surface determines the thickness of the regolith. Where the former has outpaced the latter, the mantle is thick, but where the converse has occurred, the front is exposed as a rock pediment.

The mantled and rock forms are genetically related. Mantled pediments are surfaces of transportation, i.e. surfaces shaped by erosion and deposition (redistribution) by wash and by rills, but in contrast with the pediments of sedimentary regions the mantle of debris developed on many granitic pediments is discontinuous: running water in the form of wash and rills plus a few minor streams has planed the surface of the regolith to give the smooth, mantled surfaces, and has also scoured the mantle to expose the bedrock in rock platforms.

Though there are a few exceptions, pediments are essentially of limited extent. Some rock pediments occur in isolation, but most granite pediments are fringing forms and extend at most only a few kilometres from the mountain front. Pediments are particularly well-developed on granitic rocks. There are several possible reasons for this. First, pediments are well-developed in weak rocks, and granite is particularly vulnerable to moisture attack; in any granitic massif located at or near the land surface there are large compartments of weathered material vulnerable to planation. Second, granite typically weathers to sand or grus, which is readily transported by rills and streams and spread evenly to give the smooth surface characteristic of pediments. Third, because of the low permeability of fresh granite, the weathering front is usually sharp, so that distinctive platforms or rock pediments are readily initiated and exposed. Finally, because of the marked contrast between massive and well-fractured compartments and because the latter are defined by fractures, the piedmont angle is better developed in granitic than in many other lithological environments.

4.2.3 *Relationship between pediment and peneplain*

The fringing mantled pediments that are developed around such inselbergs as Ucontitchie Hill, Eyre Peninsula, South Australia, merge without topographic break at their lower extremities with the rolling surface of the peneplain (Fig. 4.2b). Both forms are developing concurrently. Moreover, there are enough boulder clusters and other low residuals to suggest that when the latter are eventually eliminated the pediments that now surround them will extend and coalesce in convex crests that are integral parts of the peneplain.

Thus, rock and mantled pediments are not genetically or temporally distinct from peneplains. They commonly coexist. They are particular types of plains developed on fresh rock and in the piedmont of inselbergs. Pediments are not fundamentally distinct from peneplains: they are parts of a larger whole, namely planation surfaces. Moreover, though pediments are well and widely developed and preserved in arid and semi-arid lands, they occur also in other climatic contexts. Pediments are typical of the arid and semi-arid tropics but they are not, as Blackwelder (1931, p. 138) suggested, merely "the desert-inhabiting species of the genus peneplain" for both pediments and peneplains are developed on granitic rocks in various climatic contexts. Peneplains are dissected pediments and reflect the concomitant downstream increase in volume of run-off and evolution of the drainage network, resulting in the concentration of flow in fewer larger channels. These cause dissection and the conversion of the surface from smooth to undulating or rolling, depending on stream spacing.

4.3 ETCH PLAINS IN GRANITE

Many deep weathering profiles developed on granitic rocks are reported from the various shield lands which have been relatively stable for long periods. Thus it is not surprising that plains are well and widely developed on older granite masses. High plains are also characteristic of several Palaeozoic uplands, such as Dartmoor, southwestern England, the Meseta of the Iberian Peninsula and the Central Massif of France, which are, however, not of simple epigene origin but are due to a more complex sequence of events.

Where the weathered mantle developed on a surface of low relief has been stripped away usually by rivers but also by other epigene processes, the weathering front has been exposed as an etch plain or platform. That exposed near Platja d'Aro, Girona, in northeastern Spain, provides a good example (Fig. 4.9). The platform is cut in granite and its level is coincident with that of the weathering front preserved beneath the soil-regolith cover. The platform has been exposed as a result of the stripping of the regolith by waves during the last eustatic rise of sea level (Twidale, Bourne and Twidale, 1977). But etch planation is not necessarily an indication of such baselevel movement for rivers may develop lower gradients in debris reduced in size by weathering and may thus erode into the regolith. Also, lowering could result from the subsurface flushing and evacuation in solution of the products of weathering, as suggested by Ruxton (1958) and Trendall (1962).

Weathering and stripping are in some regions incomplete, as for example in the Ávila-Villacastín region of central Spain. There, the landscape is characterised by plains carrying a discontinuous recently developed regolith, with numerous boulders, clusters of boulders and even small nubbins and koppies. Remnants of an older regolith are preserved in low mesas and plateaux which stand a few metres above the general plain level (Fig. 4.10). A similar partly-stripped etch surface with many boulders (corestones) has been exposed just east of O Cadramón, in northeastern Galicia,

Plains – the expected granite form 73

Figure 4.9. Etch surface in granodiorite, Platja d'Aro, Girona, northeastern Spain. The residual surface has been stripped of regolithic cover by waves to expose the weathering front. r: regolith; g: granite.

Figure 4.10. Etch surface in granite, Ávila-Villacastín area of central Spain. The low plateau is underlain by a regolith which has been partly stripped to expose the weathering front as a bouldery granite surface.

Figure 4.11. The Labrador high plain of eastern Canada, eroded in granite and exposed by the glacial stripping of the regolith.

Spain. Most of the soil has been removed by anthropogenic action and taken elsewhere so as to improve pasture at a new site.

Much of the Labrador Peninsula, Canada, is a high plain eroded largely in granitic materials (Fig. 4.11). It has been modified by Late Cainozoic ice sheets and, in particular, has been stripped

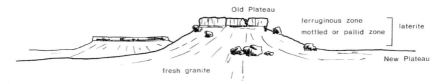

Figure 4.12.   Old and New plateaux, Cue, Western Australia.

of most of its regolith, so that it is in part at least of etch character. In addition, however, it may in part be exhumed. The extensive plains cut in granite in the southwest of Western Australia are also of etch character. The lower surface has been called the New Plateau, though it is more appropriately called a high plain. Nonetheless, there are remnants of an older lateritised plain preserved as plateaux, mesas and buttes and standing above the level of the New surface (Fig. 4.12). In many areas the lateritic weathering profile is preserved in these residuals and well-exposed in the bounding scarps that delimit them. Elsewhere, the former regolithic cover was capped by silcrete. Whatever its nature, the base of the regolith is roughly coincident with the level of the New Plateau, so that the latter is of etch character.

4.4   VERY FLAT PLAINS

Apart from rolling surfaces of low relief eroded in granite, there are, in various parts of the world, remarkably flat, featureless and extensive plains, some of which are cut in granite rocks. Several of them have been called pediplains which King (1942) saw as the end stage of a cycle of pediplanation:

> "During maturity of the landscape opposing scarps meet from the opposite sides of hills, which are thereby rapidly lowered. The residual uplands disappear, relief decreases markedly and with coalescing of the ever-increasing pediments a bevelled landscape of low relief and of multiconcave form, a pediplain ... is produced". (King, 1942, p. 53).

King cited the Springbok Flats, north of Pretoria, South Africa, as an example of such an extreme stage of pediplanation (Fig. 4.13). Many such planation surfaces, however, are really very low relief peneplains, because there is some relief, and the broad divides are convex. Even the Springbok Flats and other plains in southern Africa considered by King (1942) to be pediplains display some relief, and in particular broad convexities.

Nevertheless, planation surfaces of virtually nil relief are developed in granitic terrains in various parts of the world. Morphologically, they differ from peneplains and warrant separate consideration. The Bushman (or Bushmanland) Surface of the Western Cape Province and adjacent areas are a plain of regional extent and extreme flatness that is eroded in granite, gneiss and sandstone (Fig. 1.2e), and the plains around Meekatharra, in central Western Australia, eroded in Archaean migmatites (Twidale, 1983), are similar (Fig. 4.14). Morphologically these plains are comparable only to the incredibly flat and featureless depositional plains of central Australia (Twidale, 1981), and elsewhere. Yet they are manifestly of erosional origin, for intrinsically fresh bedrock occurs only a few centimetres beneath the surface.

The very flat plains have commonly been considered to be the end product of scarp retreat and pedimentation, and to result from the coalescence of pediments. But this suggestion is difficult to sustain as a general argument. It has been shown that scarps retreat only in certain structural circumstances. Pediments are not necessarily associated with backwearing, for some scarps are essentially fixed in space and yet pediments are associated with them. Theoretically, pediplains consist of innumerable concave-upward or rectilinear pediments, and the very flat plains under review do not conform with such a morphology.

Plains – the expected granite form 75

Figure 4.13. The Springbok Flats, a subdued but gently rolling plain, north of Pretoria and in Northern Province in South Africa.

Figure 4.14. Very flat plains in granite and gneiss, Meekatharra, Western Australia.

These flat plains are either etch surfaces reduced to extraordinary low relief by long continued subsurface moisture attack, or they represent the ultimate stage of planation, the possibility of which was appreciated by Davis (1909) and by Peel (1941). Such flat plains are most likely to evolve in areas of tectonic stability and to be preserved in areas of arid climate. The examples cited are located in arid and semi-arid areas of Africa and Australia (Williams, 1969) and in shield areas that have suffered little tectonic disturbance since the middle Proterozoic or even earlier, for over wide areas of both of these continents Upper, and even Middle, Proterozoic strata remain essentially undisturbed.

4.5 MULTICYCLIC AND STEPPED ASSEMBLAGES

Multicyclic forms are developed in granitic terrain as a result of the relative lowering of baselevel, stream rejuvenation and landscape revival. The plains become high plains, located high in the relief, as in the Sierra Nevada, in the Rocky Mountains of Colorado and Wyoming (the Sherman Surface) (Mackin, 1937, 1970); on Dartmoor; in the Ávila-Villacastín region of central Spain and throughout in the northwest of the Iberian Peninsula; in several parts of southern Africa and elsewhere.

76 *Landforms and Geology of Granite Terrains*

But they are dissected by streams working toward their new baselevel. Valley-in-valley forms are first developed (Fig. 4.15). These comprise valley-side facets separated by breaks of slope, and corresponding graded stream sectors separated by nick points in the shape of waterfalls or rapids.

In granitic terrains such breaks of slope may develop for structural reasons. Thus, particularly massive blocks may form local baselevels to which stream sectors are graded. Similarly, and especially in areas of gneissic rocks, particularly massive bands of rock may give rise to stepped relief on the valley-side slopes, as, for instance, in the Rooifontein Valley of central Namaqualand, where many such local baselevels can be related to structural factors. In some areas, however, a similar morphology is developed on the opposite side of the valley (Fig. 4.16), suggesting that the forms may be cyclic, depending on the distribution of the local (structural) baselevel. In like fashion, if waterfalls are cyclic rather than structural, they ought to be developed on all rivers in a given region, whereas structurally determined features tend to be random or isolated.

Valley-side facets tend to extend laterally and nick points inland so that, in time, a new surface of low relief comes to replace the former plain. Remnants of the latter, and indeed of even earlier

Figure 4.15. Valley-side facets (X and Y) in granite at Reedy Creek, western Murray Plains, South Australia. By correlation with the nearby Murray valley, the upper valley is of Pliocene age (Twidale, Lindsay and Bourne, 1978).

Figure 4.16. Valley-side facets or steps, stripped of regolith, and probably held up by resistant bands in the gneiss, Rooiberg, northern Namaqualand, Cape Province, South Africa.

Plains – the expected granite form 77

surfaces, may persist in the landscape. The latter has a stepped appearance and can be described as multicyclic since there is in it evidence of more than a geomorphic cycle.

In some areas, such as Cameroon, in West Africa (Mayer, 1995), stepped topography is evidently associated with faulting. Three major high plains have been recognised as well as several minor ones. The slopes separating adjacent surfaces have long been recognised and variously interpreted as fault scarps, or as due to lithological contrast and differential erosion. An interpretation rather different from this and the multicyclic concept is due to Wahrhaftig (1965) with respect to the southern Sierra Nevada, in California. Planation surfaces are a prominent feature in many parts of the upland (Fig. 4.17).

According to Wahrhaftig (1965) the stepped topography is not only confined to granitic outcrops but is found wherever granite is exposed in the region. He explains the forms in terms of the behaviour of granite in wet and dry environments:

> "The stepped topography is believed to be caused by differences in the rate of weathering in the two environments to which granitic rocks in the Sierra Nevada are subject. Where buried by overburden or gruss, the solid granitic rocks are moist most of the year, and disintegrate comparatively rapidly ... where exposed, the solid granitic rocks dry after each rain and therefore weather slowly." (Wahrhaftig, 1965, p. 1166).

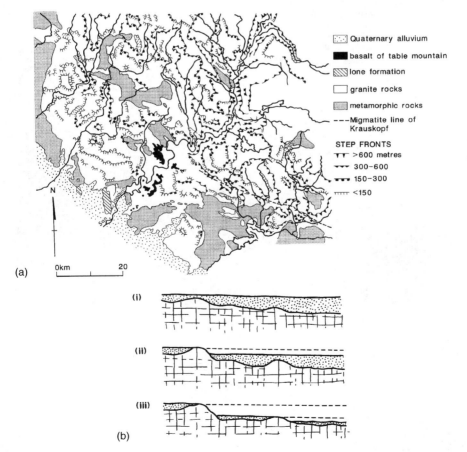

Figure 4.17. Stepped topography, Sierra Nevada, California: (a) Map (Wahrhaftig, 1965); (b) sections showing suggested mode of origin.

78 *Landforms and Geology of Granite Terrains*

Dry granite is stable and resistant whereas granite in contact with water is rapidly altered. Thus, blocks exposed in linear zones tend to become upstanding and to form local baselevels for upstream sectors of rivers crossing the barriers. In this way a stepped topography evolves, dependent in the first instance on the varied behaviour of granite to weathering. Wahrhaftig (1965) discusses various possible reasons for the upstanding outcrops becoming upstanding in the first instance, but fracture density is almost certainly the cause – the ridges are based on outcrops of massive blocks.

An alternative explanation of stepped topography can be suggested, using the scarp-foot weathering and phased or episodic exposure responsible for stepped inselbergs (see Chapters 6 and 8).

4.6 EXHUMED PLAINS

As well as being important components of modern scenery granitic plains have also been recognised in unconformity, in the stratigraphic column. Some have been exhumed to form part of the contemporary landscape. Thus, as an example of the still-buried types, in the south area of Cape Town, South Africa, the Table Mountain Sandstone rests on a planate surface cut in Precambrian granite (Toit, du, 1937). The unconformity is well exposed in coastal sections and road cuttings, and is indeed followed over considerable sectors by the main road linking Cape Town with the Cape of Good Hope (Fig. 4.18a). An unconformity of quite remarkable smoothness and eroded in considerable measure in granite is exposed in the Grand Canyon in the western USA (Sharp, 1940). A Precambrian peneplain developed in crystalline rocks and partially exhumed from beneath

Figure 4.18. (a) Unconformity between Table Mountain Sandstone and granite, south of Cape Town, South Africa. (b) Exhumed Precambrian surface of low relief inland from Kap Ingersoll, north Greenland (Cowie, 1961).

Proterozoic sediments (Ambrose, 1964) is reported from north Greenland (Cowie, 1960, Fig. 4.18b). Similar surfaces of low relief preserved in crystalline terrain, but resurrected from beneath Proterozoic and earlier Palaeozoic cover, occur in several parts of the Canadian Arctic.

4.7 SUMMARY

Plains are widely developed in granitic rocks. Fringing the uplands, pediments are well represented, but they do not extend more than a few kilometres from the higher ground. The extensive plains are either broadly rolling or of extraordinary flatness. Some of these planation surfaces are epigene in origin but the very flat ones may be of etch type, and result from long periods of subsurface weathering. They are integral components of inselberg landscapes.

REFERENCES

Ambrose, J.W. 1964. Exhumed palaeoplains of the Precambrian Shield of North America. American *Journal of Science* 262: 817–857.
Blackwelder, E. 1931. Desert plains. *Journal of Geology* 39: 133–140.
Bourne, J.A., Twidale, C.R. and Smith, D.M. 1974. The Corrobinnie Depression, Eyre Peninsula, South Australia. *Transactions of the Royal Society of South Australia* 98: 139–152.
Bryan, K. 1922. Erosion and sedimentation in the Papago Country, Arizona. *United States Geological Survey Bulletin* 730: 19–90.
Bryan, K. 1936. The formation of pediments. 16th International Geological Congress (1933) 2: 765–775.
Cowie, J.W. 1960. Contributions to the geology of north Greenland. *Meddelelser om Gronland* 164.
Davis, S.N. 1963. Silica in streams and groundwater. *American Journal of Science* 262: 870–891.
Davis, W.M. 1899. The geographical cycle. *Geographical Journal* 14: 481–504.
Davis, W.M. 1909. Geographical Essays. Dover, Boston.
Howard, A.D. 1942. Pediment passes and the pediment problem. *Journal of Geomorphology* 5: 3–31, 95–136.
King, L.C. 1942. South African Scenery. Oliver and Boyd, Edinburgh.
King, L.C.1949. The pediment landform: some current problems. *Geological Magazine* 86: 245–250.
Livingstone, D.A. 1963. Chemical composition of rivers and lakes. United States Geological Survey Professional Paper 440-G.
Mabbutt, J.A. 1966. Mantle-controlled planation of pediments. *American Journal of Science* 264: 78–91.
Mabbutt, J.A. 1978. Lessons from pediments. In Davies J.L. and Williams M.A.J. (Eds), Landform Evolution in Australasia. Australian National University Press, Canberra. pp. 331–347.
Mackin, J.H. 1937. Erosional history of the Big Horn Basin, Wyoming. *Geological Society of America Bulletin* 48, 815–860.
Mackin, J.H. 1970. Origin of pediments in the western United States. In Pesci M. (Ed.), Problems of Relief Planation. Akadémiai Kiado, Budapest. pp. 85–105
Mayer, R.E. 1995. L'origine de l'étagement de surfaces étagés abordée par la carte topographiques et des images LANDSAT: cas du socle précambrien du Cameroun septentrional. *Zeitschrift für Geomorphologie* 39: 293–311.
McGee, W.J. 1897. Sheetflood erosion. *Geological Society of America Bulletin* 8: 87–112.
Passarge, S. 1904. Rumpfläche und Inselberge. *Zeitschrift Deutsche Geologische Gesellschaft* 56: 195–203.
Peel, R.F. 1941. Denudation landforms of the central Libyan Desert. *Journal of Geomorphology* 4: 3–23.
Pugh, J. 1956. Fringing pediments and marginal depressions in the inselberg landscape of Nigeria. *Transactions and Papers of the Institute of British Geographers* 22: 15–31.
Ruxton, B.P. 1958. Weathering and subsurface erosion in granite at the piedmont angle, Balos, Sudan. *Geological Magazine* 45: 353–377.
Sharp, R.P. 1940. Ep-Archaean and Ep-Algonkian erosion surfaces, Grand Canyon, Arizona. *Bulletin of the Geological Society of America* 51: 1235–1270.
Toit, A.L. du, 1937. Geology of South Africa. Oliver and Boyd, Edinburgh.
Trendall, A.F. 1962. The formation of apparent peneplains by a process of combined lateritisation and surface wash. *Zeitschrift für Geomorphologie* 6: 183–197.
Twidale, C.R. 1981. Origins and environments of pediments. *Journal of the Geological Society of Australia* 28: 423–434.

Twidale, C.R. 1983. Pediments, peneplains and ultiplains. *Revue de Géomorphologie Dynamique* 32: 1–35.

Twidale, C.R., Bourne, J.A. and Twidale, N. 1977. Shore platforms and sealevel changes in the Gulfs region of South Australia. *Transactions of the Royal Society of South Australia* 101: 63–74.

Twidale, C.R., Lindsay, J.M. and Bourne, J.A. 1978. Age and origin of the Murray River and Gorge in South Australia. *Proceedings of the Royal Society of Victoria* 90: 27–42.

Wahrhaftig, C. 1965. Stepped topography of the southern Sierra Nevada, California. *Geological Society of America Bulletin* 76: 1165–1190.

Williams, G.E. 1969. Characteristics and origin of a Precambrian pediment. *Journal of Geology* 77: 183–207.

5

Boulders as examples of two-stage forms

5.1 THE TWO-STAGE OR ETCHING MECHANISM

As explained in the previous chapter, many granite plains, both regional and local, are weathering fronts that have been exposed as a result of the stripping of the regolith. They have evolved in two stages and are of etch type. Many other landforms, large and small, and many of them developed in granitic rocks, are of similar origin. Bornhardts and basins, for example, have developed as a result of differential weathering at the weathering front. Some gutters and flutings are due to differential weathering and erosion, again at the weathering front (see Chapters 6, 8 and 9).

Now and in the past, where meteoric waters infiltrate to the base of the regolith they exploit a wide range of structural contrasts and weaknesses – mineral banding and foliation, veins, sills and dikes, crystal boundaries and cleavage, as well as discontinuities of various origins and patterns (Vidal Romaní, 1990 and see Fig. 5.1). Some essentially linear, but in detail irregular,

Figure 5.1. Quartz veins in granite (a) in Namaqualand (Western Cape Province), South Africa; (b) in the French Pyrenees.

82 *Landforms and Geology of Granite Terrains*

Figure 5.1. (c) In a granodiorite from Panticosa, Spanish Pyrenees. (d) Foliation in granite exploited by weathering to give a faint ribbed effect, in Namaqualand (Western Cape Province), South Africa.

clefts lacking continuous detectable fractures (see Chapter 9) may reflect the preferential weathering of linear zones of crystals in strain, and the hill top (Kubicek and Migin, 1993) reported from southern Poland and elsewhere may reflect the exploitation of tensional fractures in the crests of antiformal structures. Many of these structural features are of great antiquity, so that the resultant etch forms have their origins in the distant geological past (Twidale and Vidal Romaní, 1994).

The weathering front may be continuous and reasonably planate but is in places irregular and locally discrete (Ruxton and Berry, 1957). This last is the case with one of the most common and characteristic granite forms, the boulder.

Figure 5.2. The Leviathan, a large residual boulder from northeastern Victoria, Australia (Geological Survey of Victoria).

5.2 BOULDERS – MORPHOLOGY AND OCCURRENCES

Standing either in isolation or in groups or clusters (Fig. 1.2a), boulders are, apart from plains, perhaps the most common of all granite landforms (Twidale, 1978). They are certainly the most numerous and widely distributed of the positive relief forms developed on granite. In Australia, they range in diameter from about 25 cm, to 11 m in the Devil's Marbles in the central part of the Northern Territory, and some 33 m in the huge ovoid boulder known as The Leviathan, located in the Mt Buffalo complex of northeastern Victoria (Fig. 5.2). No doubt even larger boulders are developed elsewhere. The most common diameter, however, is between one and two metres. Boulders vary in shape from spherical to ellipsoidal, and also in the degree of roundness attained. Some are virtually perfect spheres, others less rounded, others sligthly elongate and yet others are almost cubic, only the corners and edges of the original blocks being rounded (Fig. 5.3). Boulders are found in many, if not all, climatic regimes, including recently deglaciated regions in northwestern Britain, Canada, the Snowy Mountains of southeastern Australia, and in the Hesperian Massif (Iberia).

5.3 SUBSURFACE EXPLOITATION OF ORTHOGONAL FRACTURE SYSTEMS

Granites are characteristically well-jointed and are in particular subdivided into cubic or quadrangular or parallelepipedic blocks by orthogonal fracture systems. In many parts of the world spheroidal masses of intrinsically fresh rock set in a matrix of weathered rock are exposed in cuttings, quarries and natural cliffs (Fig. 5.4). They are called the heart of the block, the kernel, floater, boule, rock-kernel, a block of hard granite, core-boulder, core-stone or corestone. Each is formed within a joint block, and is surrounded by a mass of weathered granite or grus (also gruss, jabre, xábrego, sauló, saibro, arène). As weathering progresses, the outline of the fresh rock mass changes from angular to rounded because of what MacCulloch (1814, pp. 71–72) described as *"more rapid disintegration at the angles than at the sides"*, or as the same author succinctly expressed it *"Nature mutat quadrata rotundis."* (MacCulloch, 1814, p. 76). The abrupt contact

84 *Landforms and Geology of Granite Terrains*

Figure 5.3. (a) Well-rounded boulder, Devil's Marbles, Northern Territory. (b) Slightly less well-rounded boulder, Herbert River Falls area, north Queensland.

between the still solid rounded mass and the weathered material or regolith is the weathering front. In this instance, it is roughly spherical or ellipsoidal in shape rather than essentially planate, but it is nevertheless the limit of weathering and hence is the weathering front.

In most instances, the corestones are of the same composition as were the rest of the host blocks; they are not inherently more resistant than the marginal areas. That the complex of corestones and grus is *in situ* is demonstrated by the presence of veins traversing the entire rock mass (Fig. 5.5). Corestones of intrinsically fresh granite set in a matrix of weathered, friable rock are examples of spheroidal weathering (Ollier, 1971). They also represent the first stage in the development of boulders.

Boulders as examples of two-stage forms 85

(c)

Figure 5.3. (c) Subrounded blocks near Harare, Zimbabwe.

(a)

(b)

Figure 5.4. Corestone set in grus and exposed (a) in quarry on Karimun Island, western Indonesian Archipelago. The core-boulder closest to the camera has well defined marginal laminae or thin shells; (b) in the Snowy Mountains of New South Wales.

86 *Landforms and Geology of Granite Terrains*

Figure 5.4. Corestone set in grus and exposed (c) Near Tampin, West Malaysia; (d) in the recently excavated margins of the highway Lugo-Madrid near Lugo, Galicia northwestern Spain.

Their essentially near-surface occurrence, and a concomitant though not invariable diminution in degree of rounding and increase in size of spheroidal masses with depth, suggest that most such boulders are related to processes active at or near the land surface rather than to primary magmatic or to hydrothermal effects, and are due to descending meteoric waters.

That they are due to differential weathering attack beneath the land surface was recognized as long ago as 1791 by Hassenfratz (1791) (Fig. 5.6a see also 5.6b), who, commenting on exposures he had observed in the southern Massif Central, wrote:

> "...on aperçoit tous les intermédiares entre un bloc de granit dur contenu & enchassé dans la masse totale du granit friable & un bloc entièrement degagé." (Hassenfratz, 1791, p. 101).

Figure 5.5. The presence of intrusive veins traversing slopes (a) in the Rocky Mountains, near Boulder, Colorado; (b) near Pine Creek, Northern Territory.

Later work, based on observations in many parts of the world, has confirmed that many granite boulders develop in two stages. Meteoric waters, charged with gases, chemicals and biota, percolate down and along joints. The rock immediately adjacent to the fractures is attacked by the moisture (Lagasquie, 1978). Water reacts with micas and feldspars to produce clays. Solution, hydration and hydrolysis are active. Water takes quartz (silica) into solution for though fragments persist in the regolith they too eventually disappear (Trudinger and Swaine, 1979). Bacteria with a known preference for quartz or kaolinite may facilitate entry of water. The processes are relatively slow though perhaps not as slow as it has been supposed, with micas altered in decades, feldspars in centuries or

88   *Landforms and Geology of Granite Terrains*

Figure 5.5.   (c) In Sardinia. R.F. Peel (In Twidale 1978), demonstrates beyond doubt that the corestones are *in situ* and not transported.

Figure 5.6.   This is probably the site, near Chazeirolettes, in the southern Massif Central of France, where in 1791 the corestones and boulders suggested to Hassenfratz the two-stage concept of boulder formation.

millenia. That silica from these minerals and quartz goes into solution is demonstrated by the formation of small opaline and kaolinitic speleothems in many granitic terrains (see Chapter 10).

Fracture-controlled subsurface weathering transforms an essentially homogeneous rock mass into two contrasted types of material, namely the corestones of fresh rock and the grus matrix (Fig. 5.6b). Thus, the friable grus is readily washed (or, more rarely, blown) away, and the land surface is lowered. On steep slopes the grus falls away under gravity. Most of the corestones are too massive to be moved and are left *in situ*, though in some areas some of the constituent boulders, lacking support, tumble down to form a chaotic mass of boulders known in France as compayrés (Fig. 5.7). But

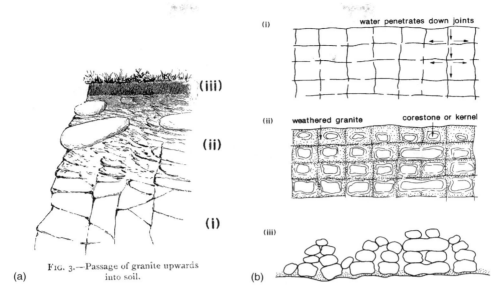

Figure 5.7. (a) Geikie's (1894, p. 15) sketch showing corestones set in weathered granite: (i) solid granite, (ii) corestones in grus, (iii) soil, (b) The two-stage development of boulders: (i) differential fracture-controlled weathering beneath the surface, (ii) fracture pattern in section, (iii) differential erosion of the differentiated mass, leaving the corestones as boulders.

Figure 5.8. Chaotic mass of boulders, or compayrés, strewn over hillslope and crest of hill at Palmer, eastern Mt Lofty Ranges, South Australia.

whatever the precise result, the exposure of the corestones by differential erosion is the second stage in the development of boulders.

In an immediate sense, two distinct processes, weathering and erosion, are involved in the formation of boulders and together they are frequently referred to as the two-stage process or mechanism (Fig. 5.8), though the two are not necessarily separate and distinct in time. The first

90 *Landforms and Geology of Granite Terrains*

Figure 5.9. Corestones in a matrix of grus, Snowy Mountains, New South Wales.

stage refers to the period of differential fracture-controlled subsurface weathering, the second to the phase of differential erosion during which grus is evacuated and the corestones are exposed as boulders. If erosion outpaces weathering, the regolith is eroded, with only a few erstwhile corestones remaining as boulders. If, however, weathering proceeds more rapidly than erosion, most corestones located in the near surface zone are reduced to grus. Those that persist emerge at the surface as boulders. Once exposed they are no longer in constant contact with moisture, and become stable, while weathering continues in the regolith, leading to the common situation of boulders being underlain by considerable thicknesses of grus (Figs 5.4 and 5.9).

The progress of weathering, and hence the shape and size of corestones, depends on the duration of uninterrupted subsurface weathering, and on the changing physics, chemistry, biology and circulation of the shallow groundwaters. The characteristics of the latter may, in some measure, reflect atmospheric climatic conditions, but they affect the rate of activity rather than the eventual

Boulders as examples of two-stage forms 91

Figure 5.10. (a) Spalling in granodiorite, near the Tooma dam site, Snowy Mountains, New South Wales. (b) Detail of part of Figure 5.10a (X).

results of weathering. Because shallow groundwaters are ubiquitous, so is subsurface weathering. Boulders of two-stage type are found wherever fracture patterns are suitable. They are widely distributed, and are developed in many and varied climatic conditions, as well as in several types of bedrock (Chapter 12).

Though conveniently termed two-stage forms, earlier magmatic, thermal and tectonic events find expression in the detailed morphology of boulders (Gagny and Cottard, 1980) (Fig. 5.5). Events that

92 *Landforms and Geology of Granite Terrains*

(a)

(b)

Figure 5.11. (a) The Peyro Clabado is a well-known perched block in the Sidobre of southern France; note also the plinth on which it sits – see Chapter 9. (b) Perched blocks, Texas Canyon, Arizona (Courtesy of J.E. Mueller).

Figure 5.11. (c) Balanced Rock, Llano of central Texas; this rock was too finely balanced for it has now been dislodged by vandals.

took place in distant geological times are frequently in evidence, so that in reality boulders are an example of multi-stage development. Thus, the shape and size of corestones and boulders reflects the pattern of weathering, which in turn and in considerable measure was determined by the geometry of the fracture pattern. This in turn is due to the stress to which the rock was subjected either during or after its emplacement, as well as the rheology of the rock, i.e. its response to stress, whether brittle, plastic, etc. Thus, crustal conditions and events of the distant past find expression in weathering patterns and the shape and size of contemporary landscape features.

This point is well illustrated by the fields of boulders exposed at Mt Monster, in the South East district of South Australia; and similar features can be seen in the Spanish Pyrenees, in the Panticosa and Cavallers districts. At these sites the orthogonal fractures that determined the present landforms were geometrically different from the original system of partings. The latter were sealed or welded by hydrothermal fluids and minerals and are now represented only by veins and veneers (at Mt Monster of quartz and plagioclase feldspar). The present fracture patterns cut across them, and it is these, together with the duration and intensity of subsurface weathering, that have determined the shape and size of corestones and boulders. The original fracture systems have not influenced weathering, erosion and landform development because of the intervention of a magmatic event.

Variations in the shape and size of blocks are determined initially by the fracture pattern, but time is also a significant factor for the extent of weathering is reflected in the size of the corestones and boulders. Cubic blocks give rise to spheroids. Elongation along one axis produces quadrangular blocks which weather to triaxial ellipsoids. Horizontal elongation produces cheese-wrings, which are so called because of their resemblance to flattish rounds of cheese. Some blocks and boulders remain essentially *in situ*, and stand either in isolation or precariously perched on other blocks or on platforms (Linton, 1955) (Fig. 5.11). Others have tumbled on to lower slopes or plains. Boulders were known as logging stones to early British workers (Fig. 5.11d), the term being derived from the verb to log or to rock; and the term has its equivalent in the penas abaladoiras of Galicia, NW Spain and Portugal. Certainly, many blocks and boulders are so finely balanced that they rock at a touch. They are also referred to as loganstones, balancing rocks, balanced

94 *Landforms and Geology of Granite Terrains*

Figure 5.11. (d) A balanced rock or logging stone (with pseudobedding) on eastern Dartmoor, southwestern England. (e) Cottage loaf at the Devil's Marbles, Northern Territory. (f) Balancing Rock, near Harare, Zimbabwe.

rocks or perched blocks. Cottage loaves (moletes in Galicia, NW Spain), comprise two or more boulders, perched one on the other, because of their similarity to old-fashioned multi-tiered bread loaves (Fig. 5.11e). Gneissic foliation gives rise to slabs, penitent rocks, monkstones, tombstones or Bussersteine (Ackermann, 1962) (Fig. 5.12a). Boulders said to resemble particular people, animals or objects are not uncommon (Figs 5.12b and c).

The size of the boulders has a genetic implication, for a corestone (and hence boulder) can be no larger than the original joint block, so that the diameter and spacing of juxtaposed corestones provide an indication of the original orthogonal joint spacing (see Chapter 6, section 4).

Figure 5.12. (a) Penitent rocks at Tungkillo, eastern Mt Lofty Ranges, South Australia. (b) Bowerman's Nose, a curiously shaped assemblage of blocks on eastern Dartmoor, southwestern England.

5.4 TECTONIC AND STRUCTURAL FORMS

Traditionally, in the English-speaking world at least, the term structural has carried two different geomorphological connotations. The phrase structural landform *sensu lato*, or used in a broad sense, encompasses both tectonic landforms due wholly to crustal activities, and structural landforms *sensu stricto*, or in the narrow sense, which are due to the exploitation of crustal weaknesses by external agencies of weathering and erosion. Tectonic forms are due to activities within the

Figure 5.12. (c) The Kennedy Rock, Matopos Hills, Zimbabwe (Lister, 1987).

crust, whereas structural forms are due to the exploitation of passive crustal features. Fault scarps and grabens are tectonic forms, in contrast with fault-line scarps and fault-line valleys which are due to the preferential weathering and erosion of zones of weakness, namely fault zones.

That the word structural has two different meanings is confusing. In addition, the definitions are oversimple and incomplete. Though some well-known granite forms are tectonic in the terms outlined above, others are more complex, and are due to crustal activities which, however, are initially suppressed by lithostatic loading, and only find expression as a result of erosion of some, at least, of the superincumbent load. Yet, others are structural in the older sense of being due to the exploitation of bedrock weaknesses, many of which, however, are due to past magmatic, thermal or tectonic events, a point made most forcibly by Lagasquie (1978). Thus, A-tents (see Chapter 11) are associated with the release of compressive stress. Weathering and erosion play no part in their development. They can be regarded as tectonic forms.

Figure 5.13. Shear planes with conjugate system of discontinuities and corestones exposed in road cutting at A Coruña, Galicia, NW Spain.

On the other hand, it has been suggested that sheet fractures and structures are due to shortening/compression (Chapter 2), implicitly acting on brittle stage at some depth in the crust. The resultant zones of strain were planes of weakness, for some were intruded by veins and sills, but they were not manifested as fractures until erosional unloading had diminished the lithostatic load. The orthogonal fracture systems that are the basis of many bornhardts (Chapter 6), as well as boulders, are similarly initiated by stress at depth, but do not find expression as partings until lowering of the land bring them close to the surface. Thus, sheet and orthogonal fractures can be regarded as released tectonic forms. There is an epigene or external component involved in the development of the fractures and associated forms, but it is indirect and referred.

The weathering and erosion of passive structures have produced a wide range of forms, major and minor, in granitic and in other lithological environments. Many of the weaknesses exploited, however, are due to crustal events dating from the distant past, and ranging from hydrothermal invasions, to the intrusion of stocks and the development of anatectic folds, to the development of foliation or cleavage. It is misleadingly oversimple to refer to associated forms as structural without reference to the crustal activities in which they are based. They can be regarded as exploited crustal features, magmatic, thermal or tectonic (see Figs 5.10 and 5.13).

Corestone boulders are examples of exploited tectonic features. In most instances their shape and size vary with fracture patterns though in some magmatic processes resulting in mineral banding (see below), for example, have played a part. These bedrock properties did not, however, find expression until erosion brought the country rock near the surface and into the zone of shallow groundwaters.

5.5 TYPES OF PERIPHERAL OR MARGINAL WEATHERING

Several variations in the type of weathering developed marginal to corestones have been noted in the field, though whatever the details, the end product is still referred to as spheroidal weathering.

Many corestones display pitting (Twidale and Bourne, 1976) (Fig. 3.5), but this probably developed subsequent to the formation of the corestones, when lowering of the land surface brought them into the near surface zone of intense weathering associated with abundance of moisture and biota. Some corestones are set in a mass of grus. Such weathering is known as granular disintegration. Some corestones are surrounded by layer upon layer of thin (1–5 mm) discontinuous flakes, slivers or laminae, wrapped around the corestones, like the leaves of a book (Fig. 3.1a).

At other sites the concentric layers are thicker (10–30 cm) and look like the leaves of an onion: hence the name onion-skin weathering, which is also referred to as spalling (Figs 3.1b and c, and 5.10). Finally, in some localities corestones have evidently been formed through the separation, by fracturing, of tetrahedral masses, each with a concave inner face, at each corner of the joint block. Some of the roughly ovoid core masses have flat ends and look like barrels (see Chapter 6, section 3.3).

Whatever the type of marginal weathering, however, almost all observers agree that the transition from the fresh rock preserved in the corestone to the friable, altered marginal areas – the weathering front – is remarkably abrupt. This sharp change is almost certainly due to the physical character of granite which is of very low porosity (Mabbutt, 1961) and permeability when fresh and cohesive, but which becomes much more permeable once it is even slightly weathered.

5.6 CAUSES OF PERIPHERAL WEATHERING

Some workers have invoked insolation as the cause of disintegration and spalling, but corestones set in grus are commonly found scores of metres beneath the land surface, far beyond the effects of diurnal, annual or even secular temperature changes. Heating and cooling cannot explain such weathering deep beneath the surface (Farmin, 1937). Surface flaking (Fig. 5.14), which is such a notable feature of granitic outcrops in arid and semi-arid lands, may, however, be of this origin, though it can also be construed as having been initiated at the weathering front and then been exposed by the stripping of all but the basal flakes of the regolith (Chapter 3). The common occurrence of concentrations of iron oxides in the laminae marginal to corestones (Fig. 3.6 pp. 57) strengthens this suggestion.

Several workers have suggested that the concentric structure observed in the marginal areas of many granite joint blocks is due to the release of pressure consequent upon erosional offloading

Figure 5.14. Surface flaking on granite platform, western Pilbara, Western Australia.

(see Chapter 2). It has been argued that during cooling and crystallisation the granite becomes stabilised in conditions of high lithostatic pressure, if only because of the weight of superincumbent rocks. Some granitic masses may have been emplaced at comparatively shallow depths, but even so, that the overlying rocks have been eroded away is evidenced by the very exposure of the granite, so that the vertical loading has undoubtedly decreased in time. The argument is persuasive and both laboratory work and practical experience in deep mines (Leeman, 1962), suggest that fractures develop parallel to the surface of voids under conditions of diminished lithostatic pressure. Thus, in deep tunnels banks of fractures aligned parallel to the voids, and therefore of arcuate shape, develop in the intradosal zones adjacent to the tunnels (Fig. 5.15) as a result of expansion of the bedrock.

Against the suggestion of pressure release are the occurrence of concentric structures around corestones in such rocks as basalts that have never been deeply buried, the occurrence of the structure all round the corestone rather than preferentially on the upper sides, and the development of flaking on the interiors of tafoni located within orthogonal joint blocks and sheet structure, in granite, volcanic rocks such as trachyte, and sedimentary rocks such as quartzite. Also, in granites and other plutonic and metamorphic rocks any tendency to expansion through unloading would, surely, be counterbalanced, in part at least, by contraction on cooling (see also Chapter 2).

Some argue that corestones reflect primary petrological structures. Nodular or concentric structures were noted in the Dartmoor granites of southwestern England and some have attributed the formation of boulders to curved joints or fissures. Curved joints certainly exist (Fig. 5.16), for example in Remarkable Rocks, Kangaroo Island, South Australia or the granites of the Serra de Xurés in southern Galicia, but they are uncommon and they have not influenced the shape of the vast majority of boulders; primary sets of concentric or spherical fractures of a radius consistent with the observed size range of corestones and boulders have so far not been located. On the other hand, it is apparent that flow in magmas has determined the distribution of various minerals and hence fractures. Thus, at many sites in the vicinity of A Coruña, northwestern Spain, fractures in the granite exposures frequently run parallel to bands of biotite. Mineral banding, a primary petrogenic feature, has also contributed to the development of corestones by influencing the course of weathering within joint blocks. Near the Tooma Dam Site in the Snowy Mountains, New South Wales, for example, mineral banding occurs in the marginal areas of blocks of diorite, and the shape and size of the corestones are clearly related to these (Fig. 5.10). Again, in the Lake Tchad region of central Africa, corestones of granite embedded in rhyolite have been explained as a magmatic feature, with

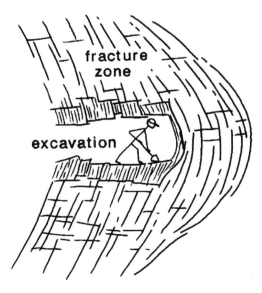

Figure 5.15. Section through intradosal zone of tunnel (Leeman, 1962).

100 *Landforms and Geology of Granite Terrains*

(a)

(b)

Figure 5.16. (a) Curved joint, Remarkable Rocks, Kangaroo Island, South Australia. (b) Curved joint, Serra do Xurés, southern Galicia, northwestern Spain.

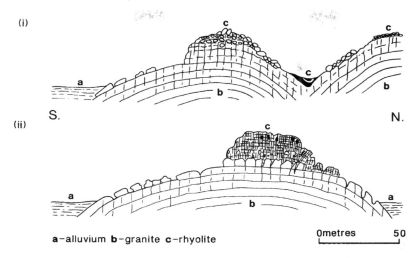

Figure 5.17. Sections through inselbergs at Ngoura and Gamsous, Lake Tchad region, central Africa (Barbeau and Gèze, 1957).

globules of still liquid granite having been mixed with the faster crystallising, and more easily weathered, rhyolite to form corestones and, in due course, boulders (Fig. 5.17).

The preferential weathering of the marginal zones has also been attributed to the presence of microfissures, which may be due to pressures generated during shearing along pre-existing fractures. Similarly, the tetrahedral cornerstones described from some few sites cannot be explained in terms of insolation, pressure release or chemical attack, but are comprehensible in terms of the rotational shearing of pre-existing orthogonal joint blocks. The elongate barrels could have developed within sheared joint blocks. Such a mechanism accounts for the observed forms, the parallelism between the long axes of the corestones and the regional tectonic style, and the notable absence of chemical alteration in the fracture zones. Similar torsional strains in the context of the pressure and temperature environments (i.e. rheology) at the time of stress could explain the contrast between flaking and spalling. Alternatively, such differences are explicable in terms of weathering, the critical factors being rock composition, and the type and degree of alteration required to cause expansion and rupture.

The marginal rotting of joint blocks in the southwest of England has been attributed to ascending hot fluids and gases penetrating along joints and effecting hydrothermal metamorphism. The survival of compartments of fresh rock above rotted rock has been cited as evidence of such an origin, but the distribution of weathering is as readily explained in terms of resistance to downward percolating meteoric waters which infiltrate laterally along fracture planes as well as vertically (Fig. 5.18). Hydrothermal intervention ought to be indicated by the occurrence of such characteristic minerals as epidote and fluorite.

Some workers consider that weathering may cause volume increase, and this has been used to explain flaking, spalling, etc. As water penetrates along the joints volume increase consequent upon alteration could cause the affected outer zone to separate from the main or host mass. As water penetrates further into each block, so more and more shells could be developed, but why some shells are thin (flakes) and others several centimetres thick (spalls) is not clear, though the contrast presumably reflects the inherent tensile strength of the rock and the amount of volume increase (assuming there is an increase) induced by weathering.

But fresh granite is a remarkably strong rock. It consists of interlocking crystals and is additionally strengthened by intercrystalline ionic bonding. Like all rocks, granite has a high compressive strength (up to 5000–6000 kg/cm^2), but even the tensile strength of unweathered granite and gneiss is high, attaining 1000–1500 kg/cm^2. Yet, the laminae involved in some of the flaking around corestones display at most slight alteration of the feldspars and biotite, and it is difficult to visualise how such slight chemical alteration and production of hydrophilic clays could cause

102 *Landforms and Geology of Granite Terrains*

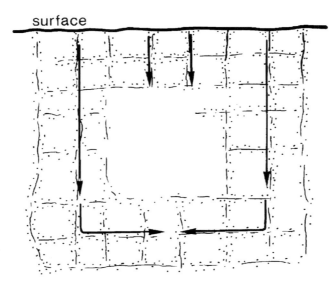

Figure 5.18. Diagram showing lateral infiltration of groundwaters to produce weathering below fresh compartment.

the rock to rupture, unless the adsorption of layered water has played a crucial part. Nevertheless, many writers have, perhaps reluctantly, felt compelled to accept such minor alteration as an adequate causation; possibly because, though unsatisfactory, there is no other obvious or more compelling explanation. Thus, Anderson (1931, p. 59) attributed the weathering of granite in Idaho to moisture attack, and while acknowledging that the *"degree of decomposition appears to be slight"*, considered that it was sufficient to cause disaggregation. Again, Larsen (1948, p. 115), working in southern California, stated that *"A slight hydration of biotite and other minerals is probably sufficient to effect the change in volume that produces the disintegration and formation of boulders."*

If volume increase has taken place, and since various minerals are involved, the expansion ought to be differential and cause disruption of textures in the rock. Yet such features as mineral banding, lineation and foliation remain undisturbed. Also, if water penetration is the cause of fracturing in the marginal zones, the flakes ought, in detail, to conform to crystal boundaries, whereas in fact they cut across such boundaries, as well as cleavage.

It is for these reasons that some workers favour an origin of concentric structure involving no volume change. They suggest that chemical reactions of the Liesegang ring type, involving diffusion and periodic reprecipitation and resolution of salts, presumably causing fatigue and disintegration, may be responsible for the observed evidence. Another possible reason for the minimal alteration noted in some of the thoroughly disaggregated granite rocks is that hydration shattering involving adsorption of ordered water has occurred. Certainly, such a mechanism accounts for the observed facts, though of course once a rock has suffered disaggregation water can more readily penetrate the mass and effect solution, hydration and hydrolysis.

Finally, it may be mentioned that though corestones are in time rotted through entirely and replaced by grus or clay, signs of some remain in the form of arcuate and concentric lines of resistant or distinctive minerals developed during weathering. Such remnants of former corestones are known as *"ghosts"*.

5.7 EVACUATION OF GRUS

Whatever the morphology, composition and genesis of the marginal zones, the weathered granite may eventually be evacuated and the corestones exposed as boulders. The transportation

Figure 5.19. Blocky corestone morainic accumulation on the margin of Touça glacier valley (Serra do Gerês, northern Portugal).

of the weathered rock is largely the work of wash, rills and rivers. A spectacular example of the concentration of residual boulders by running water occurs in the high valley of the Touça River, in northern Portugal, where glacial meltwaters have evacuated grus leaving a lateral moraine formed by a train of large boulders on the side of the river channel over a length of some 10 km (Fig. 5.19).

But wind-driven waves have achieved exposure of corestones by the preferential stripping of grus, so that entire beaches are composed of corestones, as for example on the Costa da Morte, in western Galicia, between Cape Vilano and Camelle, Spain and in Cape Willoughby, Kangaroo Island, South Australia, where the sea has stripped the regolith to expose an etch surface in granite, and the released corestones have been washed into a bouldery shingle beach or coido (see also Chapter 12). Wind may play some small part in the exposure of corestones in arid and semi-arid regions, and solifluction is significant in nival areas, but it is running water that is primarily responsible for the exposure of corestones as boulders.

5.8 BOULDERS OF EPIGENE ORIGIN

Not all boulders are of subterranean origin. Some are glacial erratics, some result from frost action, and yet others result from the disintegration of massive arcuate sheets or slabs which are typical of bornhardts or domical hills (Barbeau and Gèze, 1957 and see Chapters 2 and 6). These arcuate masses of rock are split by steeply inclined radial fracture planes (cross or fan joints). They effectively subdivide the thick slabs into blocks, which can be seen *in situ* in orderly arrangement on the flanks of many bornhardts and nubbins (see Chapter 7).

Some sheets disintegrate while still beneath the land surface and the blocks are converted to corestones set in grus (Fig. 5.20). Others, however, have been rounded and yet others completely broken down by epigene or subaerial agencies, i.e. by processes active at the land surface. There are many variations in the pattern of weathering, but in some places, for instance in the Monte Pindo of Galicia, Spain and in Corsica, France, the sheet remnants preserved on upper slopes consist of clusters of angular blocks, many with tafoni well-developed, whereas those resting on

104 *Landforms and Geology of Granite Terrains*

Figure 5.20. Blocky corestones (X) set in grus and derived from subsurface weathering of sheet structure, Paarlberg Quarry, near Cape Town, South Africa.

Figure 5.21. Disintegrated sheet structure (a) on midslope at Enchanted Rock, Llano, central Texas.

lower slopes are rounded. Such variations presumably reflect the availability of moisture at different topographic levels. Comparatively few remnants remain on midslopes, where gradients are sufficiently steep to cause blocks to slide down to lower levels, and those that remain consist of disintegrated sheets *in situ* (Klaer, 1956) (Fig. 5.21).

Boulders as examples of two-stage forms 105

Figure 5.21. (b) On a bornhardt near Garies, Namaqualand, Western Cape Province South Africa; (c) on Little Wudinna Hill, northwestern Eyre Peninsula, South Australia.

5.9 SUMMARY

Though granite boulders are formed in several ways, most are due to a combination of two processes involving, first, differential fracture-controlled subsurface moisture weathering which produces corestones set in a matrix of grus, and then erosion, most commonly by wash and streams, which evacuates the grus and exposes the corestones as boulders. Such a two-stage development clearly applies to granite boulders in many parts of the world, and has operated under

Figure 5.21. (d) On Houlderoo Rocks, southern piedmont of the Gawler Ranges, South Australia.

various and varied climatic conditions. The concept, which can be traced back almost two hundred years, is now widely accepted as a rational and reasonable explanation for these, the most common of all residual forms developed on granitic rocks.

Boulders are not climatic indicators, as assumed by some earlier workers, who mistook them for glacial erratics. Some, of course, are of this nature, in the sense that they have been transported by glaciers, possibly being further rounded in the process. But most granite boulders are residual in that they remain after the grus and other weathered granite that originally enveloped them have been evacuated. Boulders can be considered to be convergent forms (or forms of equifinality), for though most are the result of two-stage development involving fracture-controlled subsurface weathering followed by differential erosion of the unevenly weathered mass, the precise nature of the weathering and erosional processes has varied from place to place and no doubt also from time to time at the same site. Moreover, some are formed by subaerial or epigene processes.

REFERENCES

Ackermann, E. 1962. Busssersteine-Zeugen Vorzeitlicher Grundwasserschwankungen. *Zeitschrift für Geomorphologie* 6: 148–182.
Anderson, A.L. 1931. Geology and mineral resources of Cassia County, Idaho. Idaho Bureau of Mines and Geology Bulletin 14.
Barbeau, J. and Gèze, B. 1957. Les coupoles granitiques et rhyolitiques de la région de Fort-Lamy (Tschad). *Comptes Rendus Sommaire et Bulletin, Societé Géologique de France* (Series 6) 7: 341–351.
Farmin, R. 1937. Hypogene exfoliation of rock masses. *Journal of Geology* 45: 625–635.
Gagny, C. and Cottard, F. 1980. Proposition de signes conventionnels pour la représentation de certaines structures magmatiques acquises au cours de la mise en place et la cristallisation. Comptes Rendus 105ième Congrès National Societé de Savants (Caen), *Sciences* 2: 37–50.
Geikie, A. 1894. A Class-book of Geology. Macmillan, London.
Hassenfratz, J.-H. 1791. Sur l'arrangement de plusieurs gros blocs que l'on observe dans les montagnes. *Annales de Chimie* 11: 95–107.
Klaer, W. 1956. Verwitterungsformen in Granit auf Korsika. Petermmanns Geographische Mitteilungen Ergänzungsheft 261.

Kubicek, P. and Migin, P. 1993. Granite hill-top depressions in the Zulovska Pahorkatina (hilly land). *Scripta* 22: 55–60.

Lagasquie, J.-J. 1978. Relations entre les modelés d'érosion différentielle et la structure de quelques ensembles de granitoides des Pyrénées centrales et orientales. *Revue de Géologie Dynamique et de Géographie Physique* 20, 219–234.

Lagasquie, J.-J. 1984. Géomorphologie des Granites. Les Massifs Granitiques de la Moitié Orientale des Pyrenées Françaises. Editions de la Centre Nationale de la Recherche Scientifique.

Larsen, E.S. 1948. Batholith and associated rocks of Corona, Elsinore and San Luis Rey quadrangles, southern California. Geological Society of America Memoir 29.

Leeman, E.F. 1962. Rock bursts in South African gold mines. New Scientist 16, 79–82.

Linton, D.L. 1955. The problem of tors. *Geographical Journal* 121: 470–487.

Mabbutt J.A. 1961. "Basal surface" or "weathering front". *Proceedings of the Geologists' Association of London* 72: 357–358.

MacCulloch, J. 1814. On the granite tors of Cornwall. *Transactions of the Geological Society* 2: 66–78.

Ollier C.D. 1971. Causes of spheroidal weathering. *Earth Science Reviews* 7: 127–141.

Ruxton, B.P. and Berry, L.R. 1957. Weathering of granite and associated erosional features in Hong Kong. *Geological Society of America Bulletin* 68: 1263–1292.

Trudinger, P.A. and Swaine, D.J. 1979. Biogeochemical Recycling of Mineral-forming Elements. Elsevier, Amsterdam.

Twidale C.R. 1978. Early explanations of granite boulders. *Revue de Géomorphologie Dynamique* 27, 133–142.

Twidale, C.R. and Bourne, J.A. 1976. Origin and significance of pitting on granite rocks. *Zeitschrift für Geomorphologie* 20: 405–416.

Twidale, C.R. and Vidal Romaní, J.R. 1994. On the multistage development of etch forms. *Geomorphology* 11: 107–124.

Vidal Romaní, J.R. 1990. Minor forms in granitic rocks: a record of their deformative history. *Cuadernos do Laboratorio Xeolóxico de Laxe* 15: 317–328.

6

Inselbergs and bornhardts

6.1 DEFINITIONS AND TERMINOLOGY

Inselbergs are ranges, ridges and isolated hills that stand abruptly from the surrounding plains, like islands from the sea (Figs 1.2e, 1.3a and b, and 6.1a). Early English-speaking explorers referred to them as island mounts (Toit, du 1937; Willis, 1934), but the comparable German term, Inselberg, in other languages with the upper case initial letter dropped, has gained general currency in the technical literature (Bornhardt, 1900; Jessen, 1936). Some of these residuals are spectacular features and whether built of granite or of some other rock type, those located in desert and semi-desert regions, where they can be seen from afar and in toto, have a dramatic visual impact. Thus, so erudite and civilised a person as van der Post (1958, pp. 181–182) could write of the gneissic Tsodilo or Slippery Hills of northern Namibia that they rose "sheer out of the flat plain, and were from the base up made entirely of stone, and this alone, in a world of deep sand, gave them a sense of mystery". Tall, steep-sided hills like the Pão de Açucar of Rio de Janeiro, Brazil, are no less stupendous, for though there are other prominent domical forms nearby, their very size and abrupt

Figure 6.1. (a) Literal as well as littoral inselberg: some of the Cíes Islands, western Galicia, at the mouth of Vigo Ría showing marked asymmetry with cliffed slopes facing west, the dominant wind and forced wave direction.

Figure 6.1. (b) The Groot Spitzkoppe, central Namibia.

Figure 6.2. Bald bornhardts and inselbergs (a) Mt. Lindsay, Mann Range, central Australia (Mines and Energy, South Australia).

sidewalls render them unusual and eye-catching. It is because of the perceived dramatic quality of the residual hills that the inselbergs, and not the more extensive plains, have given their name to inselberg landscapes or Inselberglandschaften (Figs 1.2e, 1.3a and b, and 6.1b).

Inselbergs are characterised by steep bounding slopes which meet the adjacent plains in a sharp, almost angular, junction, known as the piedmont angle or nick. Inselbergs are of many shapes and sizes, but the granitic forms are of three major types. By far the most common and widely distributed is the bornhardt (Twidale, 1980; Twidale and Bourne, 1978), the domical form named after the German geologist Wilhelm Bornhardt (1900) who, late last century, explored parts of East Africa, and provided some of the most evocative descriptions and beautiful sketches of the forms and the landscape of which they are part, as well as astute analyses of their possible origins.

Bornhardts are bald, steep-sided domes (Birot, 1958) (Figs 2.1c and 6.2). The flanks of some gneissic forms display ribs or parallel clefts (Figs 6.3a and b), but the domical form is everywhere

Inselbergs and bornhardts 111

Figure 6.2. (b) Marehuru Hills, southern Zimbabwe. (c) Paarlberg, South Africa. (d) Enchanted Rock, central Texas (Kastning, 1976).

112 *Landforms and Geology of Granite Terrains*

Figure 6.3. (a) Bornhardt in foliated granitic gneiss, and with ribbed slopes, Rooiberg, Namaqualand (Western Cape Province), South Africa. (b) Dome in gneissic granite, Reynolds Range, central Australia. Note clefts in parallel with cleavage. (c) Whalebacks in granite, Devil's Marbles, Northern Territory.

prominent. The hills vary in size and shape. Some are low, elongate and elliptical in plan and are called whalebacks or dos de baleine (Fig. 6.3c). Those that are more nearly elliptical in plan and have steep bounding slopes are known as turtlebacks. A few are high, asymmetrical in profile and with little imagination reasonably called elephant rocks (dos d'elephant). Many have plan axes of similar length, approximately equal to the height of the crest above the adjacent plains and they are referred to as domes or half-oranges. The many local names – matopos, ruwares, morros, dwalas, meias laranjas, demi-oranges, moas, navas, medas, yelmos and so on – provide some indication of the wide distribution of this basic form.

Some bornhardts stand in isolation (Fig. 1.2e) and others occur in small groups (Fig. 1.2b), but in contrast with these detached forms there are ranges or massifs that comprise ordered repetitions of the domical form. Thus, the Everard Range, in the north of South Australia, and the Kamiesberge of central Namaqualand, each consists of a large number of closely juxtaposed bornhardts arranged in ordered rows (Fig. 6.4). In these landscapes each dome is developed on a joint block, but is nevertheless part of a larger massif. Thus, many inselbergs are bornhardts, but not all bornhardts are inselbergs.

Some granitic inselbergs are block- or boulder-strewn and are called nubbins or knolls, and yet others are angular and castellated and are known as castle koppies. Neither is as frequently and widely developed as the domed variety. These forms appear to be variations on the bornhardt theme,

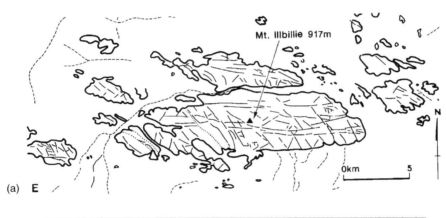

Figure 6.4. (a) Map of part of the Everard Ranges, northern South Australia (drawn from air photographs), showing orthogonal fracture system. (b) The development of a bornhardt on each fracture defined block.

114 *Landforms and Geology of Granite Terrains*

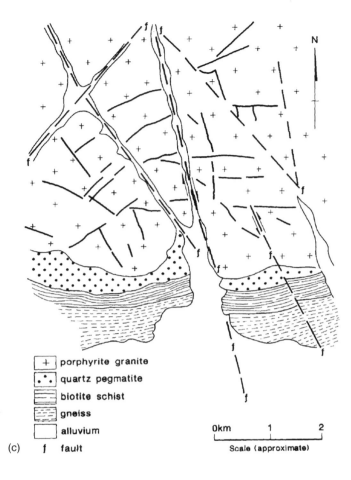

Figure 6.4. (c) Map of part of the Kamiesberge, Namaqualand (Western Cape Province) showing fracture pattern (drawn from air photographs). (d) Bornhardts developed on individual fracture-defined blocks in the area shown in Fig. 6.4c.

and for this reason, plus their frequency of occurrence and widespread distribution, bornhardts are first analysed as the basic form from which some of the others are derived. Nubbins, koppies and other uplands are discussed in Chapter 7.

6.2 BORNHARDT CHARACTERISTICS

Bornhardts are domical hills with bare rock exposed over most of the surface (Fig. 6.2). Though essentially domical, many are bevelled (Fig. 6.5). They occur in orogens as well as cratonic regions, in hilly terrains and uplands as well as lowlands and plains. They are well-developed in granite (Whitlow, 1978–9) and several other rock types, though only in those that are massive (see Chapter 12).

Bornhardts are not restricted to any particular climatic environment but occur in most climatic regimes, from arid to humid, hot to cold (Godard, Lagasquie and Lageat, 1994) (Fig. 6.6). Inselbergs are restricted neither to the tropics, nor to arid and semi-arid lands, though they are well

Figure 6.5. (a) Bevelled bornhardt in central Zimbabwe (Lister, 1987). (b) Sisarga Grande, Malpica, north-western Spain.

116 *Landforms and Geology of Granite Terrains*

Figure 6.6. Domical granitic residuals (a) in the Sahara Desert (Rognon, 1967); (b) in the humid tropics near Rio de Janeiro, southeastern Brazil, and including the Pão de Açucar (Brazilian Tourist Bureau).

represented and easily viewed in the latter areas. They are found, for example, in Finland, northern Norway (e.g. the Kråkmotinden), and the mountains of southern Poland (Migoń, 1993).

In plan form, bornhardts are delineated by prominent vertical and near-vertical fractures that form part of orthogonal systems and such partings also determine the location and development of many features within the uplands (Fig. 6.7). Arcuate convex-upward sheeting joints are coincident with domical outlines of many of the residuals (though, as is mentioned in Chapter 2, there are some curious exceptions). Also, excavations reveal that many morphologically simple domes are eroded in complex structures comprising multiple domes. Finally, bornhardts are developed in multicyclic landscapes, that is, in landscapes in which remnants or palaeoplains preserved high in the relief indicate former phases of baselevelling, subsequent relative uplift and stream incision (Fig. 6.8).

Inselbergs and bornhardts 117

Figure 6.6. (c) In the seasonally cold Domeland, central Sierra Nevada, California. Note the conifers, the set of steeply inclined fractures and also remnants of sheet structure (National Park Service).

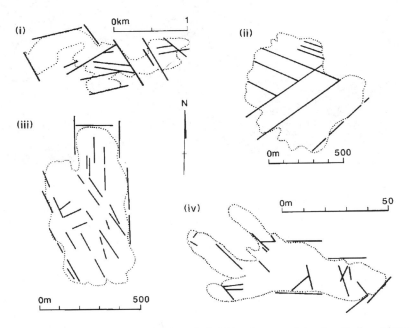

Figure 6.7. Plans of various inselbergs showing relationship of topography and vertical or subvertical fractures: (i) Hyden Rock and (ii) Kondinin Rock, both in Western Australia; (iii) Wudinna Hill and (iv) Pildappa Rock, both on Eyre Peninsula, South Australia.

Figure 6.8. (a) A multicyclic landscape near White River, in the eastern Mpumalanga South Africa. The high plain (X) is a remnant of an Early Cainozoic planation surface, with a residual hill in the Bushman's Kop. It has been dissected, exposing incipient domes or bornhardts in the scarp, and a new planation surface is forming at a lower level in the foreground. (b) Granitic hills in the valley of the Umgeni River, KwaZulu Natal, South Africa, with (African – Early-Mid Tertiary) planation surface clearly visible on the skyline.

6.3 THEORIES OF ORIGIN

6.3.1 *Environment*

As their name suggests, inselbergs, including some bornhardts, were at one time compared to rocky islands. Some few authors went further and suggested that the residuals were of marine origin. Bornhardt (1900) himself initially entertained this view, but brief reflection shows that the

Figure 6.8. (c) Remnant of planation surface in granitic terrain of southern Greenland (Oen Ing Soen).

argument cannot be sustained. Bornhardts are spectacularly developed in coastal and inshore zones in such areas as the Baia Guanabara on which stands the city of Rio de Janeiro, Brazil (see Fig. 6.6b), in northwestern Spain (see Fig. 6.5b), and on several sectors of the coasts of southern Australia and of Galicia, NW Spain (Fig. 6.1a), but they are also well represented in the continental interiors that have not been touched by the sea for scores, even hundreds, of millions of years, if ever. For example, there is no evidence that the Yilgarn Craton of Western Australia, a region rich in bornhardts, has been inundated by the oceans during the past 1,000 Ma.

In some areas marine processes have been responsible for the differential erosion that has produced the inselbergs, and for the asymmetry of some examples (e.g. Fig. 6.1a), but there is no evidence to suggest that marine forces alone are capable of producing inselbergs in general, or bornhardts in particular. It is merely that waves and other marine agencies (including biota) have exploited weaknesses in the bedrock exposed in the coastal zone, leaving resistant masses in relief.

Explanations of bornhardts based on climate have long been popular. Agassiz (1865) interpreted the granite boulders of the Rio de Janeiro area as glacial erratics and a similar construction was placed on boulders in Costa Rica. He took the granitic domes of the same area of southeastern Brazil to be huge roches moutonnées. Le Conte (1873) reached the same conclusion with respect to the domes of the Yosemite in central California, USA. In reality, there is no evidence that the Rio area has been glaciated during the relevant time period, and though the Yosemite has been affected by Pleistocene glaciers and some of the domes (e.g. Gilbert Dome) have been markedly modified by them, they appear to be structural forms which predate the appearance of the ice. The domes of the Sierra Nevada, California, and of northern Norway, occur in glacial or glaciated regions, but, as with marine environments, their climatic context appears to be incidental. The capacity of glaciers to erode appears to depend on the physical condition of the basal ice (cold, warm, frozen, wet-based). Glaciers are capable of bulldozing (Kleman, 1994) pre-existing regoliths and of exposing the weathering front. Humid-climate glaciers like those of France and Spain have not only stripped the regolith but have also destroyed any minor etch features formed at the weathering front. But there is no reason to suppose that glaciers or ice sheets are inherently capable of the differential erosion required to produce domes from homogeneous bedrock. Similarly, though freeze-thaw activity can prepare the rock for transport, and though solifluxion may achieve the evacuation of some detritus, the mechanism cannot account for the differential erosion of a homogeneous rock mass to produce bornhardts.

The exposure and unimpeded view of inselbergs in arid or semi-arid lands have a considerable psychological impact, and have persuaded several workers that there is a genetic link between landform and desert climates. Passarge (1895), for example, urged that some of the inselbergs of West Africa are the result of wind erosion of the intervening plains during the Mesozoic, and, early this century regional planation by the wind was favoured by workers such as Keyes (1912) and Jutson (1914). But even in arid lands the idea of widespread aeolian erosion is inconsistent with the field evidence, and the concept enjoyed only brief and limited favour, both generally and in respect of inselbergs. And though recent work suggests that some extensive playa depressions, for example in northern Iran, are substantially of aeolian origin, these features are excavated in weak unconsolidated materials. There is no evidence that wind can erode at a regional scale and in resistant country rock.

That the plains adjacent to or surrounding bornhardts have been eroded by rivers was suggested by Bornhardt (1900), Falconer (1911) and many others. Like Passarge (1895) they implied differential erosion but emphasised fluvial action. Rivers have undoubtedly been responsible for erosion of the regolith and for the extensive exposure of the weathering front in etch plains (Chapter 4). Thus, in specific instances exposure of the bornhardt masses can, in whole or in part, reasonably be attributed to wave action, or to glacier ice, or to nival processes, but rivers have been most widely active.

6.3.2 *The scarp retreat hypothesis*

Fluvial erosion of a particular type is invoked by many writers, and particularly by the late King (1949, 1968). They interpret inselbergs as Fernlinge, monadnocks de position, or remnants of circumdenudation remaining after long-continued scarp retreat and pedimentation (Fig. 6.9a). Vertical fractures determine the location of major rivers, and hence, indirectly, divides or interfluves, and scarp recession proceeds from the valleys thus determined.

The idea of scarp retreat can be traced back to Fisher (1866, 1872) in the mid Nineteenth Century. The mechanism was invoked by Dutton (1882) in their interpretations of the landscapes they explored in the American West, Holmes and Wray (1912) and Holmes (1918) explained some of the gneissic inselbergs of Mozambique in terms of scarp recession. But it was King (1949, 1968) who applied the concept to landscape evolution both in a broader sense and as a basis for a general theory of inselberg development. Most proponents of scarp retreat argue that the process is restricted to caprock situations in arid and semi-arid lands. This assertion is incorrect for scarp retreat occurs in any environment in a caprock situation, as for example in sedimentary terrains in central Labrador. Nevertheless, King (1949, 1968) while passionately urging that scarp retreat and pedimentation are dominant wherever running water is active (i.e. everywhere save in glacial areas and the dune deserts), nevertheless conceded that the process attains optimal effects in the semi-arid tropical and subtropical lands.

Certainly, inselbergs are well-developed and preserved in such regions. Rates of geomorphological activity are low in desert lands. Scarp-foot weathering and erosion lead to the steepening and the recession of slopes and to the formation of a pronounced piedmont angle (see Chapter 9). Furthermore, granite is resistant under dry conditions, but is very susceptible to alteration when in contact with water (Chapter 3). In hot, arid and semi-arid lands not only is the contrast in erosional vulnerability between high and low levels in the local topography more pronounced than elsewhere, but the dry granite effectively acts as a caprock so that, even in granitic terrains with no duricrust development, the formation and maintenance of escarpments is theoretically feasible.

Jessen (1936) interpreted the inselbergs of Angola in terms of scarp retreat, but suggested that the bounding slopes became steeper late in the cycle (Fig. 6.9b). This could be explained in terms of a diminishing rate of scarp recession, which, as in plateau terrains (Willis, 1936), would allow more time for basal weathering and scarps being regraded to a maximum inclination commensurate with stability.

Proponents of the scarp retreat hypothesis attribute the rounded form of bornhardts to differential weathering under epigene attack. This explanation had earlier been applied to boulders (see Chapter 5) and in respect of larger residuals by Mennell (1904) who pointed out that outstanding edges of fracture-defined blocks are readily removed. Falconer (1911, p. 246) thought that the inselbergs of Nigeria "naturally assume that configuration of surface which afforded the least scope for the activity of the agents of denudation". The domes of the southeastern Piedmont of the USA have also been construed as due to rounding by weathering (granular disintegration).

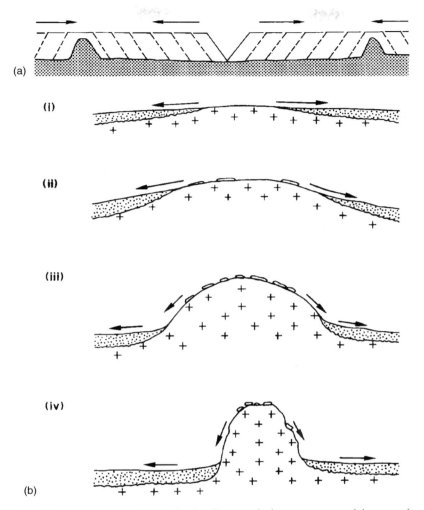

Figure 6.9. (a) The scarp retreat model showing inselbergs as the last remnants surviving on major divides. (b) Jessen's model of reduction of inselbergs by scarp retreat but with progressive steepening of slopes.

Such weathering of corners or edges of fracture-defined blocks could take place either on exposed surfaces under attack by epigene processes or beneath the surface, in the groundwater zone, as is well illustrated by the formation of corestones (Chapter 5). Those who interpret bornhardts as remnants of circumdenudation and the rounded form as due to the weathering of exposed corners and edges by epigene agencies also construe sheet fractures and structures as having formed in response to pressure release and in relation to the shape of the landform (but see Chapter 2).

6.3.3 *Tectonics and structure: faulting and lithology*

Some investigators interpret bornhardts as Härtlinge, monadnocks de résistance or de dureté, that is, as what many refer to as structural forms *sensu stricto*, or forms due to the exploitation of crustal weaknesses by external agencies. Several explanations involving the intrinsic characteristics of the country rock have been offered for bornhardts standing above the adjacent plains. Basically,

122   *Landforms and Geology of Granite Terrains*

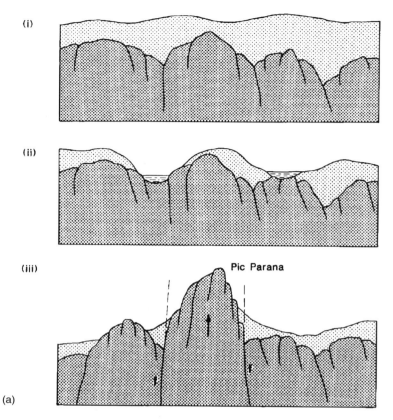

Figure 6.10.   (a) Development of the Pic Parana, southeastern Brazil, as an upfaulted hill (Barbier, 1957).

however, there are four main arguments, involving faulting, lithological control, cross-folding and contrasts in fracture density.

First, some few bornhardts are of tectonic origin for they appear to be upfaulted blocks. In this regard it is interesting to note that the local Aboriginal tribe, the Pitjandjara, believed that Ayers Rock, a well-known bevelled bornhardt of Cambrian arkosic sandstone which stands in isolation above the desert plains of central Australia (see Chapter 12), simply rose out of a large flat sand hill. There is a didactic philosophy based on myth and legend, and they sought no evidence, but had they done so they would not have found faults delineating the arkosic bornhardt (Mountford, 1965).

The Pic Parana, on the other hand, is a horst residual located in southeastern Brazil, and delimited by faults that have been recently active (Fig. 6.10a). Other bornhardts may well be of similar type, but at the Pic Parana the crucial evidence (Barbier, 1957) was exposed in excavations related to a hydroelectric scheme. Unless there are such exposures, it is difficult to demonstrate first that the residual in question is delineated by faults, and second that the faults have recently been active. Bearing in mind this proviso, the field evidence suggests that most bornhardts are not tectonic for there is no evidence that most of them are delineated by large-displacement faults, and, even in those examples cited as of tectonic origin, it is not everywhere apparent whether the fault-delineated scarps are tectonic or structural, whether they are fault or fault-line scarps. Few bornhardts are demonstrably horsts.

This is not to suggest that faults play no part in bornhardt morphology, for they form zones of weakness within residuals, for example the Pão de Açucar massif in the Rio de Janeiro area of

Inselbergs and bornhardts 123

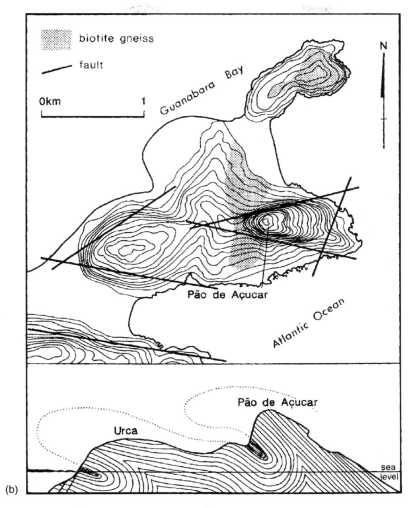

Figure 6.10. (b) Section through the Pão de Açucar, Rio de Janeiro, southeastern Brazil, showing faulting within the mass (Lamego, 1938).

southeastern Brazil (Fig. 6.10b), that have been exploited by weathering and erosion (Barbier, 1967). Again, many of the granite residuals of the Úbeda-Linares area, in Andalusia, southern Spain, which are being exhumed from beneath a Triassic cover (see below), are based on fault blocks; but the fractures are passive rather than active, and have provided avenues of weathering and erosion rather than causing the landforms to be uplifted.

Also, shearing associated to thrust faulting along subhorizontal planes, is evidenced in many granite exposures, not only by sheet structures with demonstrable dislocation but by displaced slabs and wedges (see chapters 2 and 11). It has been suggested (Vidal Romaní, 1989, 1991; Vidal Romaní and Twidale, 1998) that large scale thrusting (i.e. low angle faulting) accompanied by drag and imbrication may have caused the development of elongate ovoid masses (Fig. 6.11a) like those exposed in recent road cuttings near Lugo and in A Coruña, in northwestern Galicia (Fig. 6.11b). These masses of fresh rock (Vidal Romaní, 1989, 1991; Vidal Romaní and Twidale, 1998) associated to large-scale thrust may account for the delineation of elongated boulders of rock. As Mitra (1992) suggested for sedimentary rocks, this kind of features are common in large-scale thrust folded sequences in granitic terrains below similar deformational regimes. At the scale of blocks

124   *Landforms and Geology of Granite Terrains*

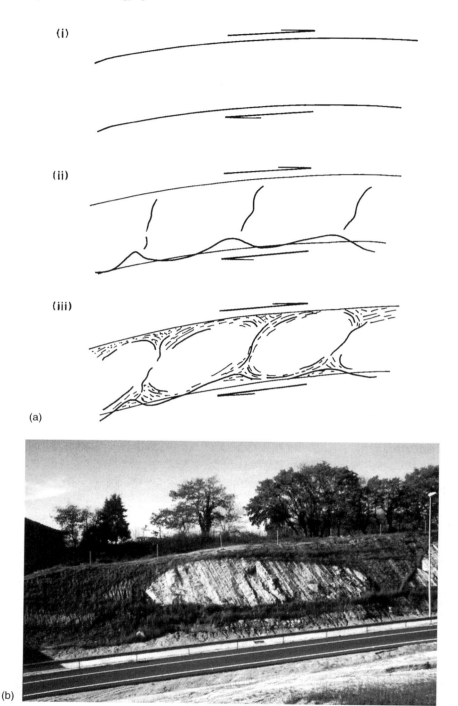

Figure 6.11. (a) Schematic development of ovoid masses of fresh rock set in foliated matrix as a result of shearing along low-planes of dislocation. (b) Highway cutting near Lugo, Galicia, northwestern Spain, showing two large corestones set in shattered and weathered matrix corresponding to a major plane of subhorizontal shearing.

and boulders, and as already described (Chapter 5), they may be bordered by laminated rock that is easily weathered below the soil. The elongated masses of fresh rock may be several tens, or even a few hundreds, of metres, in diameter (Fig. 6.11a and b), and give rise to small scale bornhardts. The mechanism could also explain the asymmetry of some domical hills such as the elephant rocks described from some parts of the world.

In this connection, it is of interest that many years ago Brajnikov (1953) suggested that some of the bornhardts (morros) of southeastern Brazil differ from other bornhardts (which are widely accepted as being huge projections of bedrock) in that they are gigantic "floaters", or detached masses of fresh rock set in a matrix of weathered material. Such detached masses of fresh bedrock could be associated with major shear zones.

On the other hand, many bornhardts appear to reflect bedrock structure and to be due to the exploitation of weaknesses in the crust by various external agents of weathering and erosion. In the Guitiriz region of northwestern Galicia, the rounded granite hills display low domical outcrops, as well as large residual boulders and small koppies. Exposures in shallow quarries suggest a direct relationship between surface form and subsurface structure, for in the granite there are developed many features due to horizontal shortening, including overthrusting and vertical wedges, as well as small, mainly large-radius, domes developed in granitic rocks (Figs 2.9a–c).

Variations in composition and texture have been neglected as causes of differential weathering and erosion within plutonic bodies. Most plutons are complex composite features within which there are several bodies of varied petrological, and hence weathering and erosional characteristics. Some of these internal variations are pronounced, others subtle. In some cases, as for instance in the exposures of Hiltaba Granite Suite of northwestern Eyre Peninsula, South Australia, there is no correlation between petrology and topography. Stone Mountain, Georgia, is a granitic dome, and some of the adjacent plains are eroded in biotite gneiss, though on the northern and eastern sides of the residual granite also underlies the plains (Fig. 6.12), suggesting that factors additional to composition are involved. But in French Guyana leucogranites form the residuals that stand above plains eroded in biotite gneiss, granitic residuals stand above plains eroded in schist in the Air Mountains of the southern Sahara, and in parts of Namibia and in southern Africa potash-rich masses tend to be upstanding. Aplogranite and fine grained granite are cited as giving rise to

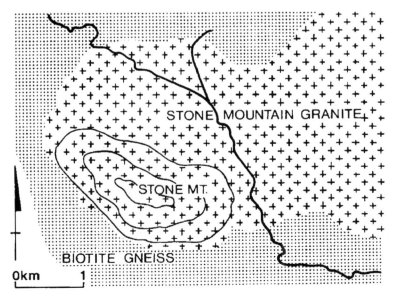

Figure 6.12. Map of Stone Mountain, Georgia, eastern USA, showing relation between topography and geology (Herrmann, 1957).

Figure 6.13. (a) An exposed stock of Donkahoek Granite intruding schist, central Namibia.

domical hills in the Sudeten Mountains of southern Poland (Migoń, 1993). In the northern Flinders Ranges, South Australia, Mt. Ward and the Armchair are pegmatitic, standing above the level of an all slopes topography eroded in a complex of granite and gneiss.

In this context, however, it must be pointed out that much of the evidence that would permit the significance of lithological contrasts to be assessed has disappeared. It is the character of the compartments of rock that were formerly located above the present plains, vis-à-vis the surviving bornhardt masses, that is crucial; and these have been eroded. All that can be done is to point to known compositional and textural variations within plutonic masses and to the possible geomorphological implications of such diversity.

It has been suggested that some of the bornhardts of Mozambique (Holmes, 1918) are merely projections or apophyses developed at the margins of the plutons, or that they are simply exposed stocks, and similar claims have been made in respect of forms in western Arabia, Zimbabwe, and Namibia (Fig. 6.13a). Again, several of the prominent domes in and around Mitchell's Nob, in the Musgrave Ranges of northern South Australia, are exposed granite stocks intruded into gneiss (Fig. 6.13b). Thus, some bornhardts are undoubtedly a manifestation of more resistant intrusive masses. In most batholithic exposures, however, there are indications of profound erosion of the crystalline rocks, suggesting that the original outlines of the plutons have long since been lost. But whatever the validity of these claims, lithological contrast is again involved.

Cross-folding may also cause the development of resistant compartments on which bornhardts are based. An analogue is provided by well-known sedimentary inselbergs from central Australia (Twidale, 1978). The plains surrounding the Olgas and Ayers Rock are eroded in sediments (conglomerate and arkose respectively) that are physically contiguous with the strata exposed in the residuals. The Olgas, Ayers Rock and Mt. Conner (a sandstone-capped mesa) are aligned E-W and the alignment cuts across the fold structures evidenced in the three residuals. Some time after a later Palaeozoic orogeny the regions suffered N-S compression resulting in cross-folding in the Early Palaeozoic strata in which the residuals are formed, and in the development of compressional cores that are the present uplands. In similar fashion, and as suggested by Dale (1923), sheet fractures delineate structural domes and basins, which argue either shearing or cross-folding (compression from markedly different directions) and resulting in an egg-box type of structure at a regional scale (Vidal Romaní, 1991). The zones of pronounced compression so developed could resist water penetration and weathering, and form the bases for residual hills. Deep erosion, taking the land surface through the upper tensional zone of antiforms and into the deeper compressional sectors, is implied.

Figure 6.13. (b) Domical hill developed on granite stock, eastern Musgrave Ranges, northern South Australia.

6.3.4 Variations in fracture density

Many of the explanations of bornhardts based on lithological variation seem to have local validity. In many other instances, however, the granite underlying the plains appears to be mineralogically similar to, if not identical with, that of which the residuals are composed. The Hiltaba Granite Suite of northwestern Eyre Peninsula, South Australia, displays several textural variations (e.g. equigranular, porphyritic), but bornhardt inselbergs are common to all of the plutons, and whatever the local petrology the same granite that is exposed in the residuals underlies the adjacent plains. The reason for contrasts in weathering and erosion in these, and in many other, areas appears to be variations in fracture density between juxtaposed compartments of rock.

Many writers have alluded to variations in fracture density as a significant factor in landform development on granitic rocks. Thus, Le Conte (1873, p. 327) noted that the massive domes of the Yosemite region of the Sierra Nevada, California, *"consist of hard material, little affected by joints"* and, more specifically, Mennell (1904), writing of the Matopos of Zimbabwe, stated:

"... the influence of the divisional plane of the rocks must not be overlooked, and it is to the variations in the number and character of the joints that the varied scenic aspects of the Matopos may be traced. Where stretches of comparatively level country occur, it will generally be found that the joints are numerous and irregular in direction, so that the rock readily breaks up and presents a large surface to the agencies of disintegration. In such cases the superior hardness of particular bands avails them little, as they are unable to show a solid front to the disrupting forces. On the other hand, joints are often entirely absent over considerable areas, and the tendency of the rock then is to weather into smooth rounded surfaces with a very large radius of curvature. Probably the actual outlines of the hills or ridges and the general direction of most of the Matopo valleys are determined by widely spaced master joints, which have been exploited by erosional agencies." (Mennell, 1904, p. 74).

6.3.5 Differential subsurface weathering and the two-stage concept

A different perspective was introduced into the bornhardt debate with the suggestion that subsurface weathering is involved and that, like many boulders (Chapter 5), many, perhaps most, bornhardts develop in two stages. One involves differential subsurface weathering, commonly controlled by variations in fracture density, but also possibly involving lithological contrasts; the other, differential erosion which exposed the bedrock projections as bornhardts (Fig. 6.14a). Rivers have most commonly been responsible for the erosional stripping of the regolith, but wind-driven waves, frost action (solifluxion) and glaciers and ice sheets (as in Scandinavia and in some degree in western Iberia) have also played their part in specific locations. Tectonic uplift may have facilitated stream erosion and the stripping did not necessarily take place in a single phase.

Though the development of bornhardts is justifiably compared to that of corestone boulders, most workers envisage that, whereas corestones become detached from the parent mass as a result of weathering, bornhardts remain contiguous with or attached to the mass of fresh rock beneath the regolith. Some pillars, intermediate in size between boulders and bornhardts, illustrate this point (Fig. 6.14b).

Again like boulders, various stages previous to differential subsurface weathering and subsequent differential erosion, and involving early magmatic, thermal and tectonic events, have to be taken into account if the forms are to be more completely understood. Although the two stages of subsurface preparation and subsequent exposure (Twidale and Bourne, 1975) are all important, the interactions of shallow groundwaters and bedrock involve many structural nuances of ancient origin, so that bornhardts, like boulders, are multi-stage features. For example, the formation of orthogonal fractures is critical to the eventual development of both boulders and bornhardts, but this occurred long before the rock mass concerned was in the groundwater zone and thus susceptible to differential weathering (the usual Stage 1). In the dacitic Gawler Ranges (Campbell and Twidale, 1991) it can be demonstrated that the orthogonal fractures had already developed in Middle Proterozoic times, some 1,400 Ma ago, whereas their exploitation by groundwaters in the shallow subsurface did not take place until the Jurassic, more than 1,200 Ma later. Yet, the formation of the orthogonal fractures in the distant geological past was critical to the eventual formation of bornhardts. Where lithological contrasts have been exploited by subsurface weathering the origin can be ultimately traced back to thermal or magmatic events of the distant past.

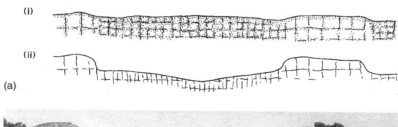

Figure 6.14. (a) The two-stage concept of bornhardt development: (i) differential fracture-controlled subsurface weathering, (ii) differential erosion of the contrasted compartments so developed. (b) Murphys Haystacks, near Streaky Bay, on northwestern Eyre Peninsula, South Australia, consists, in part, of a group of pillars still in physical continuity with the underlying granite mass. As such they can usefully be regarded as miniature bornhardts.

The gist of the two-stage hypothesis in the context of fracture density is that those compartments that are riddled with open fractures are more rapidly and intensely weathered because water can more readily penetrate the mass. They are reduced to plains underlain either by grus or by corestones set in grus. The massive compartments, on the other hand, remain essentially fresh and are resistant to erosion. The initial contrast in fracture density need not be pronounced, but once the contrast has been exploited, it is reinforced. Even at the weathering front, projections would shed water, and when exposed, the topography would be accentuated, because the residuals shed runoff and tend to remain dry, whereas the plains receive water and, thus, the rocks beneath them become more and more weathered: another example of a reinforcement or positive feedback effect.

Discussing the explanation of bornhardts based on variations in lithology, it was pointed out that the relevant evidence has gone, for it has been eroded, and the same objection can be levelled at the fracture density hypothesis. In principle it can, but in practice it has been shown that fracture patterns at the surface provide a good guide to fracture patterns and spacing at depths of a kilometre or so. This being the case, extrapolations upwards into the mass of rock removed by erosion surely also are valid? If this extrapolation is correct, the contrasts in fracture density at the surface provide a valid comparison of densities between the massive compartments that remain inselbergs and the compartments at the same topographic level that have been weathered and eroded. Thus, the suggestion that many bornhardts are coincident with relatively massive compartments appears to be soundly based. Even in areas that appear to be well fractured, like the Mt. Sinai area of the Sinai Peninsula, northeastern Egypt, domical shapes evolve if the partings are tight and impenetrable to water.

Fracture density is evidently critical to the development of some bornhardts. Several possible explanations have been suggested for such variations. According to Lamego (1938), for example, the distribution of fractures is directly related to the distribution and sense of stress and strain in folded crystalline sequences, with some zones in compression, others in tension. The deeper zones of antiforms are in compression, the shallower ones in tension and the converse applies in synformal structures (Fig. 6.15). Similar stress variations are associated with offset or en echelon transcurrent faults, and recurrent shearing and dislocation of orthogonal sets of regional magnitude could result in distortion of the cubic or rectangular blocks with a tendency to stretching along axes aligned at roughly 45° to the direction of stress. Compression and tension do not cancel out but are additive, and are of the same order of magnitude as the stresses applied. Continued dislocation causes fracture propagation in the zones adjacent to the primary fractures so that each major block comes to consist of a stressed core set in a fractured zone (Fig. 6.16).

Part of this hypothesis is susceptible to testing, for strain can be measured. This has been done on several of the Eyre Peninsula, South Australia, inselbergs, by drilling shallow holes, inserting strain gauges, and measuring the sense and amount of distortion for several hours. Results are consistent with the shear theory, with the measurements from some sites, especially in the Wudinna district, indicating contrasted stresses along different horizontal axes.

Of course, the two-stage explanation can, and ought to be, as is implied in the discussions outlined above, extended to include not only contrasts in susceptibility based in fracture density, but

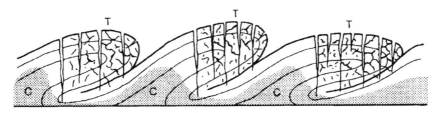

Figure 6.15. Section through the Rio de Janeiro, southeastern Brazil, area showing suggested relationship between morros or bornhardts, and antiforms and synforms developed in the crystalline rocks of the region (Lamego, 1938).

130 *Landforms and Geology of Granite Terrains*

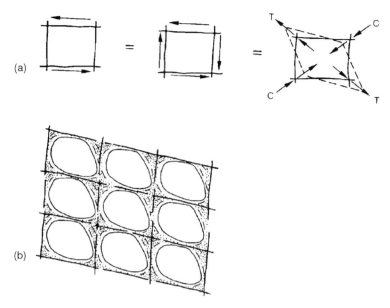

Figure 6.16. Effect of repeated shearing of cubic blocks: (a) single block, (b) compartment with several blocks (Twidale, 1980).

also those due to variations in rock composition and texture. And Falconer (1911) long ago provided an admirably succinct summary of the two-stage hypothesis embracing this broader view:

"*A plane surface of granite and gneiss subjected to long-continued weathering at base level would be decomposed to unequal depths, mainly according to the composition and texture of the various rocks. When elevation and erosion ensued, the weathered crust would be removed, and an irregular surface would be produced from which the more resistant rocks would project. Those rocks which had offered the greatest resistance to chemical weathering beneath the surface would upon exposure naturally assume that configuration of surface which afforded the least scope for the activity of the agents of denudation. In this way would arise the characteristic domes and turtlebacks which suffer further denudation only through insolation and exfoliation. Their general elliptical outlines, which Merrill (1897) would ascribe very largely to the influence of crustal stress and strain, are probably in great part due simply to the modification by weathering of original phacolitic intrusions.*" (Falconer, 1911, p. 246). And again, referring to boulder-strewn granite hills in Hausaland, he wrote that within the walls of Kano:

"*Kogon Dutsi, the larger of two flat-topped hills of diorite, although deeply decomposed, still preserves in its lower part detached boulders or cores of unweathered rock. If the subsequent erosion had continued until the weathered material had been entirely removed, the flattened hill would have been replaced by a typical kopje of loose boulders resting upon a smooth and rounded surface of rock below.*" (Falconer, 1911, p. 247).

Thus, the etch or two-stage concept appears to offer a comprehensive explanation of bornhardts and inselbergs. The exposure of the resistant compartments has most commonly involved river erosion, and it may be that where the materials being attacked were weak, slope decline most likely occurred. Where the plains between bornhardts were duricrusted, or protected by a lag accumulation, scarp retreat operated. But where it occurred it is local, ephemeral, and not an essential feature of the mechanism, merely the means by which plains have been lowered, incidentally resulting in the exposure of the inselberg residuals. Whichever mechanism was operative, however, the plains separating the residuals were lowered by rivers. But they exploited a structural contrast and were not responsible for the formation of the residuals.

6.4 EVIDENCE AND ARGUMENT CONCERNING ORIGINS OF BORNHARDTS

The scarp retreat concept not only fails to accommodate and explain several features of bornhardts, but it is also inconsistent with some of the field evidence and argument. Yet, most geologists and geomorphologists still favour scarp retreat. Why? What is the evidence and argument?

6.4.1 *Contrasts in weathering between hill and plain*

According to the fracture density hypothesis, there ought to be contrasts in density between hill and plain. Deep weathering of granite beneath plains has been reported from several parts of the world. In addition, at several sites there is a marked contrast between the depth of weathering of the bedrock beneath the plains and that of the bornhardts. Thus, at the margins of Ucontitchie Hill, Eyre Peninsula, South Australia, the granite has suffered marked differential fracture-controlled weathering, with corestones set in a matrix of weathered rock (Fig. 6.17a). Drilling shows that around the inselberg the regolith commonly extends to depths of 25 m and as deep as 40 m, yet only a few metres away, the massive rock exposed in the bornhardt remains essentially fresh. Similar contrasts in fracture density between hill and plain are evidenced near Garies, in Namaqualand (Western Cape Province) (Fig. 6.17b), near Bangalore in peninsular India (Fig. 6.17c), and on eastern Dartmoor, southwestern England (Fig. 6.17d).

6.4.2 *Incipient domes*

If bornhardts are, indeed, initiated in the subsurface by differential weathering, as it is implicit in the two-stage hypothesis, there ought to be examples of all stages of their formation, including some domical masses formed of intrinsically fresh rock, that have either just been exposed or are located immediately beneath the natural land surface. Several examples have been noted in the literature and in the landscape. From Ebaka, in south Cameroon, Boyé and Fritsch (1973) described a quarry opened to provide ballast for a railway line. The excavation exposed a domical mass of fresh granite surrounded by weathered rock. Its crest was located 8–10 m beneath the land surface, and it dipped away in all directions (Fig. 6.18). This was surely an incipient bornhardt awaiting natural exposure but revealed by artificial excavation. The dome has not been shaped by epigene processes and then buried, for the cover material is weathered granite *in situ*.

Elkington Quarry was recently (1995) opened near Pildappa Rock, north of Minnipa, on northwestern Eyre Peninsula, South Australia. Prior to excavation, only a small platform was exposed, but this proved to be the crest of a large-radius dome, with sheet structure well-developed (Fig. 6.19a). Several other low domes on northwestern Eyre Peninsula have been quarried (e.g. at Calca, south of Streaky Bay, and Quarry Hill, near Wudinna) and in each instance they have been shown to be but the crests of much larger domes, the radius of which increases with depth below the surface.

Near the Leeukop, in the northern Free State, South Africa, the crest of a dome is exposed in a shallow depression excavated as a water storage (Fig. 6.19b). In the Vredefort brick quarry, in the northeastern Free State, the weathered granite has been excavated, revealing mainly corestones but, in one corner of the quarry, part of an incipient dome is exposed (Fig. 6.19c). At Midrand, (formerly Halfway), between Johannesburg and Pretoria, South Africa, what was a small rock platform has been revealed by road excavations to be the crest of a small dome (Fig. 6.19d). Similar features have been noted at Buccleuch, south of Johannesburg; between Neue Smitsdorp and Pietersburg in the Northern Province; at the several sites between Vanrhynsdorp and Nuwerus in Namaqualand, Western Cape Province; at the Pomona Quarry, near Harare, Zimbabwe. At several sites in the southwest of Western Australia, particularly in the Darling Ranges, domes like Sullivan Rock have only just been exposed from beneath the lateritic and bauxitic regolith.

6.4.3 *Subsurface initiation of minor forms*

If minor forms associated with bornhardts are initiated beneath the land surface, it follows that the host mass must also have evolved beneath the land surface. As it is made clear in later chapters

Figure 6.17. Contrasted fracture densities between upland and plains shown in excavations at (a) Ucontitchie Hill, northwestern Eyre Peninsula, South Australia.

(8 and 9), there is irrefutable evidence that some minor features characteristic of granite domes have been initiated at the weathering front. Thus, excavations at the margins of several inselbergs on northwestern Eyre Peninsula, South Australia, show that both basins and gutters are developed at the weathering front. In their account of the exposure of the Ebaka Quarry, south Cameroon, Boyé and Fritsch (1973), record that the newly exposed domical surface is scored by gutters and basins. Here, and on Eyre Peninsula, South Australia, there is no possibility of the forms having developed subaerially and then been buried, for the overlying material is grus *in situ*.

6.4.4 *Flared slopes and stepped inselbergs*

Many of the inselbergs of southern Australia display steepened, or flared, basal slopes. There is convincing evidence that flared slopes are a particular form of the weathering front. They are initiated

Inselbergs and bornhardts 133

Figure 6.17. (b) Near Garies, Namaqualand (Western Cape Province), South Africa; (c) near Bangalore, southern India (Büdel, 1977, p. 109); (d) in and around Blackingstone Rock, eastern Dartmoor, southwestern England.

beneath the natural land surface (see Chapter 8), for at Yarwondutta Rock, near Minnipa, on northwestern Eyre Peninsula, South Australia, and elsewhere, such concavities can be seen in excavations beneath a cover of grus *in situ* (Fig. 6.20). Where such flared slopes are exposed, it is clear that the upper shoulder of the flare marks the location of a former hill-plain junction.

134 *Landforms and Geology of Granite Terrains*

Figure 6.18. Granite dome exposed by quarrying at Ebaka, south Cameroon (Boyé and Fritsch, 1973).

Figure 6.19. (a) Southern part of Elkington Rock, a large-radius dome exposed by excavation north of Minnipa, northwestern Eyre Peninsula, South Australia. (b) Crest of granitic dome (foreground) exposed near the Leeukop, a bevelled dome in the northern, Free State, South Africa.

Inselbergs and bornhardts 135

Figure 6.19. (c) The Vredefort brick pit in 1979, showing part of a granite dome (X) exposed in a corner of the quarry. (d) Granite dome exposed by roadworks at Midrand, between Pretoria and Johannesburg, South Africa.

Yarwondutta Rock provides a clear, compact example (Figs 6.21a and b). Former hill-plain junctions are indicated by the upper shoulders of the two flared slopes preserved on the northern side of the Rock. Bearing in mind the origin of flared slopes, it is deduced that:

− Yarwondutta Rock has emerged and grown as a positive relief feature as a result of the episodic or phased lowering, by streams, of the surrounding plains.
− This emergence has taken place not gradually but in distinct phases or episodes. The flared weathering fronts imply comparative standstill, time during which scarp-foot weathering has taken place. Their exposure, on the other hand, indicates stream rejuvenation and landscape revival, during which the plains were lowered, but the residuals remained in relief. Whether the scarp-foot zone was etched out deeper and deeper in successive stages of development (Fig. 6.22a), or whether there were successive lowerings of the plains (Fig. 6.22b) is not known.

136 *Landforms and Geology of Granite Terrains*

Figure 6.20. Yarwondutta Reservoir, northwestern Eyre Peninsula, South Australia, showing concave-upward weathering front exposed by the excavation of the water storage in 1915–16. The concrete pillars originally supported a corrugated iron roof intended to reduce evaporation and contamination. Note the flared basal slope in the background.

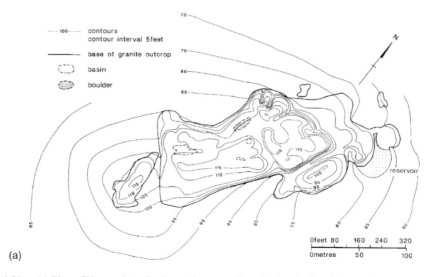

Figure 6.21. (a) Plan of Yarwondutta Rock, northwestern Eyre Peninsula, South Australia.

– Scarp-foot weathering, the first stage in the development of another set of flared forms, is in progress, for a concave weathering front is exposed in an excavation.
– Remnants of landforms related to earlier phases or cycles of development have clearly persisted. Yarwondutta Rock appears at one time to have been merely a low platform with, at most, a few boulders or sheet remnants standing above the general level. That platform is now the upper surface. It has survived at least two subsequent phases of subsurface weathering and erosion of the adjacent plain.

Figure 6.21. (b) Stepped northwestern slope at Yarwondutta Rock, northwestern Eyre Peninsula, South Australia.

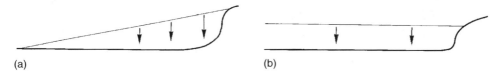

Figure 6.22. Sections showing possible mechanisms of slope lowering: (a) etching out of scarp-foot zone and (b) lowering of the whole piedmont plain.

- There has been some backwearing of the bounding slopes, not by river action for the residual is of limited extent and does not generate streams. But there is a substantial runoff in the form of wash (hence the reservoir shown in Fig. 6.20) and weathering by moisture in the scarp-foot zone has caused the bedrock slope to recede. This backwearing amounts to only a few metres, which is of quite a different order of magnitude from the scores, even hundreds of kilometres demanded by inselberg landscapes in many parts of the world and by the scarp retreat hypothesis.
- The residual has grown episodically in relief amplitude in time. This explanation meets the objection, due to King (1949, 1968) to the two-stage hypothesis of bornhardt development that the maximum depth of weathering recorded in a given region is commonly much less than the height above the plains of the highest inselbergs in that region. The implication was that some inselbergs are too high to have been initiated in the subsurface, but only a single cycle of weathering and erosion, not multicyclic or multiphase development, was considered (Twidale, 1982). Stepped inselbergs indicate episodic exposure. The cause of the implied alternations of weathering and erosion (tectonic, climatic, changes in erodibility due to weathering) remains speculative. Phases of stripping and exposure can be identified in the sedimentary records of adjacent basins.

Stepped inselbergs are not peculiar to Eyre Peninsula or even to Australia. On such bornhardts as Wudinna Hill and Ucontitchie Hill on Eyre Peninsula, and Kokerbin Hill and Hyden Rock in the southwest of Western Australia, linear subhorizontal zones of flared slopes and associated platforms, steepened slopes, breaks of slope, and tafoni occur at various levels, so that the residuals have a stepped appearance (Fig. 6.23). Moreover, inselbergs with distinct bevels and steps occur in many parts of the world, and are for example clearly illustrated from Angola in the superb sketches of Jessen (1963) (e.g. Fig. 6.24).

138 *Landforms and Geology of Granite Terrains*

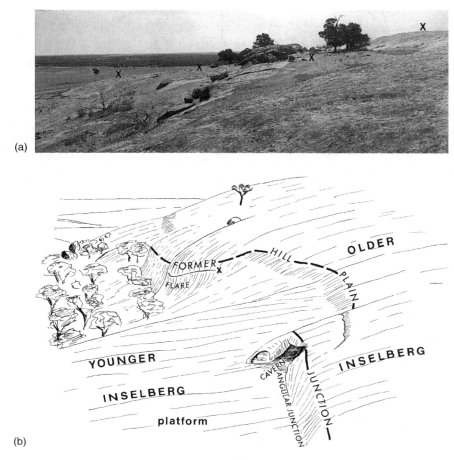

Figure 6.23. (a) Steps (X) on slope of Poondana Rock, near Minnipa, northwestern Eyre Peninsula, South Australia. (b) Stepped slope of Wudinna Hill, northwestern Eyre Peninsula, South Australia.

Figure 6.24. (a) Bongoberg and (b) Amboira, stepped inselbergs in Angola (Jessen, 1936).

6.4.5 Regional and local patterns in plan

If bornhardts and related forms are in any way due to crustal stress, they ought to be disposed in plan patterns that are geometrically conformable with regional tectonic style. They are. Thus, on northwestern Eyre Peninsula the residuals occur on NW-SE trending rises that parallel known regional fractures. In detail they display clear alignments and fracture patterns related to these broader trends (Figs 1.12, 6.25a and b). Similarly, in the Traba Massif of Galicia, NW Spain residual hills stand on fracture-defined ridges (Rodríguez, 1994) (Fig. 6.25c).

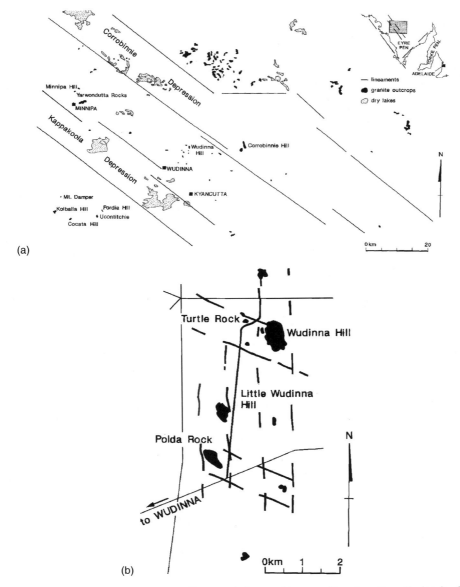

Figure 6.25. Distribution of granite inselbergs and domes: (a) on northwestern Eyre Peninsula, South Australia; (b) Wudinna Hill and adjacent areas.

Figure 6.25. (c) Distribution of residual hills in relation to fractures, Traba Massif, Galicia, northwestern Spain (Rodríguez, 1994).

6.4.6 Coexistence of forms associated with compression/shearing

Sheet fractures are everywhere associated with bornhardts. As was suggested in Chapter 2, there are grounds for interpreting them as due to crustal compression. Also, A-tents and other forms associated with the release of compressive stress (Ritchot, 1975) (Chapter 11) are widely developed on bornhardts. Thus, the coexistence of bornhardts, sheet structure and an assemblage of minor forms is consistent with the hypothesis in terms of which bornhardts are an expression of compressive stress, either directly applied or related to shear.

6.4.7 Topographic settings

Some bornhardts occur in plains settings, and, as they include some of the most spectacular examples known, they are given greater significance than they perhaps deserve. If bornhardts were the last remnants surviving after scarp retreat and pedimentation, the residuals ought to be found only in plains contexts, and they ought also to be restricted to major divides. They are not. Many residuals stand on divides, but others occur in valley floors or in valley-side slopes (Figs 6.8a and 6.26). Many occur in upland settings – in the Umgeni valley (Valley of a Thousand Hills) in Natal; in the Yosemite and Domeland, both in the Sierra Nevada of California; in the Rio de Janeiro region, Brazil; in the Kamiesberge of Namaqualand; in the Rocky Mountains of Colorado (Lester, 1938; Herrmann, 1957); and so on. These residuals surely argue against the suggestion that the forms are inherently Fernlinge, or the last remnants surviving after long-distance scarp retreat.

6.4.8 Occurrence in multicyclic landscapes

Bornhardts characteristically occur in multicyclic landscapes (Fig. 6.8). The link between bornhardts and multicyclic landscapes is twofold. First deep erosion is implied by the exposure of granitic rocks, whatever their origin, and deep differential subsurface weathering arguably requires a period or periods of standstill and baselevelling, such as is evidenced by palaeosurfaces of low relief. It is significant that most Zimbabwan inselbergs stand on a high plain located lower than the prominent African surface (Lister, 1987) (Fig. 6.27), and that in Kenya most are located on

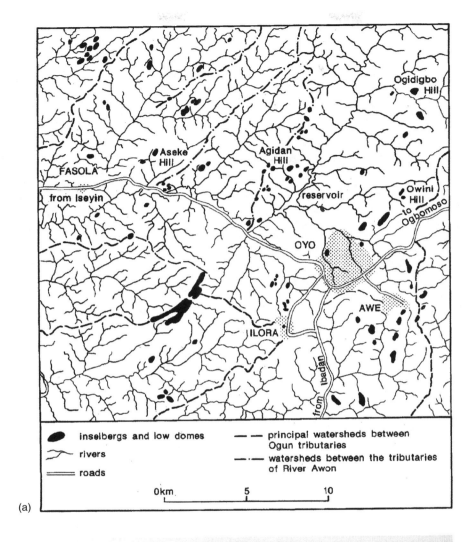

(a)

(b)

Figure 6.26. (a) Distribution of bornhardts in the Oyo region of Nigeria, showing them located on divides (Jeje, 1973). (b) Low granite dome exposed in floor of valley near Malmesbury, just north of Cape Town, Western Cape Province, South Africa.

Figure 6.26. (c) Granite domes exposed in sidewalls of glaciated valley of the Thompson River, in the eastern borders of the Rocky Mountains, near Boulder, Colorado.

Ikeda's (1991) medium elevation surface, and lower than its highest, the implication in both regions being that differential structurally-controlled subsurface weathering took place beneath a high plain. The well-known domical forms of the central Sierra Nevada, in California, are clearly located beneath and developed in association with a prominent high-level planation surface (Fig. 1.1g). In these terms the bevels that are so prominent on some bornhardts (e.g. Fig. 6.5) are construed as etch surfaces related to phases of subsurface weathering.

Second, the erosion achieved during n + 1 cycles is consistent with the suggested exposure of deeper compressional zones of the antiformal structures (Figs 6.28a and b) that become bornhardts with well-developed sheet structures but with few open orthogonal fractures. Large-radius domes with few visible fractures and few residual boulders or slabs, such as Polda Rock and Little Wudinna Hill, both in the Wudinna district of northwestern Eyre Peninsula, South Australia, are interpreted as domes located below the neutral planes of antiforms. In residuals like the nearby Ucontitchie Hill, on the other hand, with several layers of sheet structure and exposed boulders, the neutral plane may be coincident with the surface of the main mass but lie below the various sheets represented only by large blocks and boulders (Fig. 6.29).

6.4.9 Fracture-defined margins

Many bornhardts are defined by prominent vertical or near-vertical fractures (Fig. 6.7). That the margins have been stabilised on these fracture zones is indicated by the development of such features as flared slopes and scarp-foot depressions. It is surely too much of a coincidence that retreating escarpments should have been worn back to, and apparently stabilised on, these fracture zones. Surely, if long distance scarp retreat were operative, somewhere it would have regressed deeply into non-fractured rock?

6.4.10 Age of inselbergs and bornhardts

If inselbergs in general, and bornhardts in particular, are the last remnants surviving after long-continued erosion (scarp retreat), no bornhardt ought to be of an age greater than the duration of

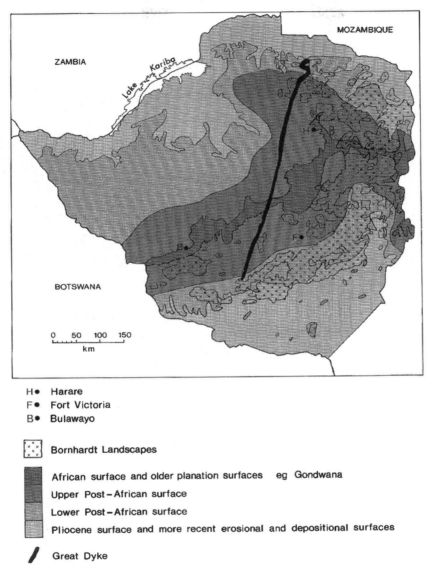

Figure 6.27. Distribution of bornhardts in respect of palaeosurfaces in Zimbabwe (Lister, 1976).

a geomorphic cycle which has attained an advanced stage, or the time taken to reduce a land mass to baselevel. If, on the other hand, bornhardts are etch forms based in structural variations, they can be of any age; provided, of course, that they can survive weathering and erosional agencies.

Estimates of the duration of a cycle vary widely. In particular, there is a major discrepancy between figures derived from fission track dating, and dates based in stratigraphy. Accepting, *pro tempore*, stratigraphic dates, there is general agreement that even after due allowance is made for isostatic recovery large areas of high land would be reduced to baselevel in periods of the order of 33–35 Ma. Thus, no inselberg ought to be of greater age than this; in stratigraphic terms no inselberg ought to predate the Oligocene, and most ought to be much younger.

144 *Landforms and Geology of Granite Terrains*

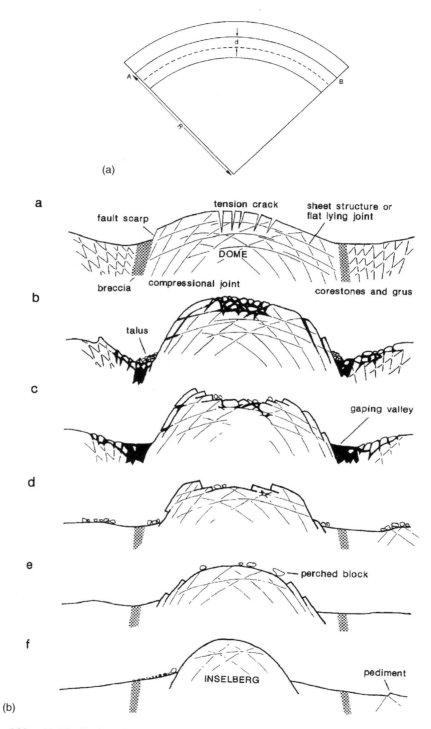

Figure 6.28. (a) Distribution of strain in an antiform. (b) Ritchot's model, showing preferential erosion of tensional crest and exposure of compressive core.

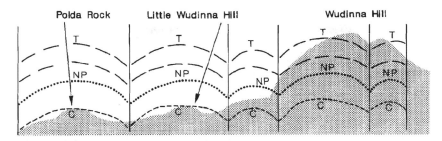

Figure 6.29. Diagrammatic sections through various inselbergs on northwestern Eyre Peninsula, South Australia, showing suggested relationship between profiles and strain.

Figure 6.30. The Humps is a group of granite gneiss domes located on the Old (lateritised) Plateau, in the southern Yilgarn of Western Australia. It and its surrounding plains are of the order of 60–100 Ma.

Bornhardt massifs like the Everard Ranges, northern South Australia, stand above plains on which silcrete developed and which are therefore probably at least of Early Cainozoic age. The crests of some of the bornhardts of northwestern and northern Eyre Peninsula, South Australia, may be of Mesozoic age, and this finds support in the putative Jurassic-Early Cretaceous age of the summit surface of the adjacent Gawler Ranges. Again, evidence has been adduced to suggest that some inselbergs preserved on the Yilgarn Block, in the southwest of Western Australia, are of late Mesozoic and earliest Tertiary age (Fig. 6.30), and some residuals on Wilsons Promontory, southern Victoria, are Early Cretaceous forms (Hill, Ollier and Joyce, 1995).

This is not to suggest that all inselbergs are ancient forms. On the contrary, many are clearly youthful, having been only recently exposed. But others are old in that they have been exposed to the elements for scores of millions of years, and they argue against scarp retreat. On the other hand, they are compatible with the two-stage hypothesis. Their survival poses problems. But it is not enough to state that they exist, therefore they must be possible. Their survival requires explanation. Several factors may be invoked. Once in positive relief, the hills shed water; they are massive with few open fractures so that most water runs off the surface; they are intrinsically dry sites, and as *"water is your sore decayer"* (Shakespeare, Hamlet, V, i, 160–161 In Hibbard (Ed.), 1994), the bedrock is altered and eroded, but only slowly. Bornhardts, like other ancient uplands, are thus best explained in terms of reinforcement or positive feedback mechanisms.

6.5 EXHUMED BORNHARDTS AND INSELBERGS

Many inselbergs and bornhardts of exhumed type are recorded in the literature and landscape. They are of various ages. Thus, the magnificent inselberg landscapes described by Bornhardt (1900) from what is now Tanzania, are exhumed and of preCretaceous age, and some of those considered by Falconer (1911) in his seminal studies of Nigerian residuals are also exhumed but of preEocene age (Figs 6.31a and b). Tall domical residuals (pains de sucre) have been exhumed from beneath the Late Palaeozoic Tassili Sandstone, in the Tassili Mountains of southern Algeria (Rognon, 1967) (Fig. 6.31c), and the partial stripping of a cover of Nama (Cambrian) strata has resurrected a field of granitic nubbins in southern Namibia and the adjacent parts of Namaqualand (Western Cape Province) (Fig. 6.31d). In Australia, exhumed granitic inselbergs range in age from Early Pleistocene (northwestern Eyre Peninsula), through pre-Miocene (western Murray Basin) to numerous examples of earliest Cretaceous or Jurassic age (Fig. 6.32) to Late Archaean in the Pilbara region of Western Australia. In Spain, the inselbergs of the Úbeda – Linares area of Andalusia, Southern Spain, predate the Triassic (Fig. 6.33a), and the exhumed granitic terrain of Charnwood, in the English Midlands, is of similar age (Watts, 1903). In northwest Scotland, the (Neoproterozoic) Torridonian Sandstone in northwest Scotland rests on what appears to be an old inselberg landscape, developed in Lewisian gneiss, and consisting of domical residuals surrounded by pediments (Fig. 6.33b). The younger sediments have been partly eroded to exhume the ancient landforms.

These examples contradict the view that bornhardts are Cainozoic forms. On the contrary, they are of ancient lineage and, whatever mechanism or mechanisms are responsible, have been feasible and active through much of geological time, and definitely for the past 2.5 Ga. Considered together, their temporal and spatial distributions suggest that they can form in a wide range of climatic conditions. Though the volume of water at and near the Earth's surface may have increased through time, some has been present throughout the geological record. Thus, there has always been water available for weathering, consistent with the requirements of the two-stage mechanism.

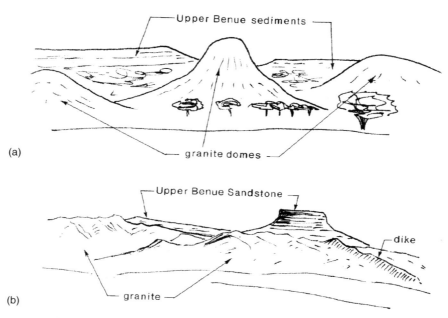

Figure 6.31. (a) and (b) Sketches drawn from photographs (Falconer, 1911) showing exhumed preEocene granitic forms in northern Nigeria.

Bornhardts could have been initiated by subsurface structurally-controlled weathering in any climatic context, though the agents responsible for burial, and for subsequent exposure, have no doubt varied from place to place and time to time. Thus, the inselbergs of Fennoscandia and Galicia-Northern Portugal were initiated in warm humid conditions during the Tertiary but were exposed by glacial action during the Quaternary, whereas the stripping of tors on Dartmoor, southwestern England, is in part at least due to nival action, and the (phased) exposure of some of the Eyre Peninsula, South Australia, inselbergs, for instance, can be attributed largely to fluvial action,

Figure 6.31. (c) Section showing granite hills re-exposed by the partial stripping of the cover of Carboniferous sandstone (and by the erosion of the grus from the fracture zones?) in the Tassili Mountains of southern Algeria (Barbier, 1967). (d) Sketch showing exhumed subCambrian granitic hills from the Namibia-Namaqualand (Western Cape Province) border district (Du Toit, 1937).

Figure 6.32. (a) These granite domes near Port Hedland, in the north of Western Australia, have been covered by the Early Cretaceous seas and sediments (the latter forming the plateaux), and then partly re-exposed. They are of exhumed type and preCretaceous age.

Figure 6.32. (b) The inselberg landscape, with low granite hills, east of Kulgera, central Australia, was inundated by Late Jurassic or early Cretaceous seas and sediments (preserved in the mesa, left) and then exhumed.

Figure 6.33. (a) Granite domes in process of exhumation from beneath: a Triassic marine cover, in the Úbeda-Linares district of southern Spain; (b) Lewisian hills (L) from beneath Torridonian (Neoproterozoic) sandstone (T) in northwest Scotland (Robinson and Williams, 1989 In Vidal Romaní and Twidale, 1998).

though volume decrease of the regolith consequent on alteration and subsurface evacuation in solution and by flushing may also have contributed.

6.6 ANTIQUITY AND INSELBERG LANDSCAPES

The quintessential and classical inselberg landscapes are developed on the shield lands of Africa, Australia and India; and for good reason, for it is in such ancient continental nucleii that there has been time for long-continued weathering, resulting in the eradication at the weathering front of all but the most durable resistant compartments. Even these, however, have been reduced to minor proportions. There has also been time for the stripping, partial or complete, of the regolith and for the exposure, again partial or complete, of the residual remnants as inselbergs or massifs, which despite being areally subordinate to the plains, are so prominent and eye-catching that they, and not the plains, have given their name to the landscape as a whole. Though high rate of activity can compensate for time, inselbergs and inselberg landscapes reflect great age and in particular long periods of subsurface weathering.

6.7 SUMMARY

Bornhardts are domes in both the structural and the topographic senses of the term. They form in a variety of ways, but most are Härtlinge, monadnocks de résistance or de dureté, which have formed in two major stages. This interpretation accounts for the field characteristics. Though initiated by differential structurally-controlled subsurface weathering, the stripping of the regolith and exposure of the weathering front is in places due to ice sheets and glaciers, elsewhere to wind-driven waves, though mostly to river action. Thus, they are convergent in the sense that similar forms have evolved in different ways.

Many appear to reflect exploitation by shallow groundwaters of variations in fracture density, though the effects of mineralogical variations have probably been underestimated. Many of the structural weaknesses exploited at the weathering front have their origins in magmatic, thermal and tectonic events of the distant past. For this reason it is suggested that, though it is convenient to consider such etch forms as having developed in two-stages, in reality granite bornhardts are multi-stage rather than two-stage in origin. Moreover, deep-seated crustal events continue to influence landform development in granitic terrains (Blès, 1986).

It has long been accepted that inselbergs and bornhardts are climatic forms; the particular climatic connection has varied from time to time and from writer to writer, though savanna or desert environments have found most favour. But groundwaters are ubiquitous, so that in terms of two-stage development the first of the requirements necessary for bornhardt evolution obtains all over the continents. Cool climate processes cause exploitation of steeply inclined fractures and produce angular forms, and warm humid conditions permit soil formation, the formation of vegetated debris slopes and a softening of the piedmont angle, in marked contrast with desert bornhardts. But the basic forms are similar, regardless of climatic setting.

Inselbergs, on the other hand, are particularly well represented in stable, shield regions, whether arid or humid, warm or cool: such forms have been reported from Fennoscandia and Newfoundland as well as Angola and Namibia, from the selvas of Nigeria (Jeje, 1973) and the monsoon lands of peninsular India, as well as the mediterranean lands of southern and southwestern Australia. Many were initiated in warm humid conditions though the agent responsible for exposure has varied. But the common and crucial factor in inselberg development is not climate but time. The shield lands are relatively stable and have long been exposed to weathering, and in particular subsurface weathering. There has been time for even the most resistant of rocks to be reduced to small size; most have been substantially weathered, and only the cores of even the toughest compartments of resistant rock survive as inselbergs surrounded by extensive plains, or as inselberg landscapes (Brook, 1978).

Thus, the two-stage mechanism, though an oversimplification of reality, is consistent with the known distribution, global, regional and local, of bornhardts. They can be expected to occur wherever and whenever structural conditions are suitable, regardless of climate, to be exposed in a range of topographic situations, in uplands and plains, on divides, in valley sides and in valley floors.

REFERENCES

Agassiz, L. 1865. On the Drift of Brazil, and on decomposed rocks under the Drift. *American Journal of Science and Arts* 40: 389–390.

Barbier, R. 1957. Aménagements hydroélectriques dans le sud du Brésil. *Comptes Rendus Sommaire et Bulletin Societé Géologique de France* 6: 877–892.

Barbier, R. 1967. Nouvelles réflexions sur le problème des 'pains de sucre' à propos d'observations dans le Tassili N'Ajjer (Algérie). Faculté des Sciences de Grenoble. *Travaux Laboratoire de Géologie* 43: 15–21.

Birot, P. 1958. Les domes crystallins. Centre National de la Recherche Scientifique *Memoirs et Documents*, 7–34.

Blès, J.L. 1986. Fracturation profonde des massifs rocheuses granitiques. Bureau de Recherches Géologiques et Minières Document 102.

Bornhardt, W. 1900. Zur Oberflächengestaltung und Geologie Deutsch Ostafrikas. Reimer, Berlin.

Boyé, M. and Fritsch, P. 1973. Dégagement artificiel d'un dome crystallin au Sud-Cameroun. *Travaux et Documents de Géographie Tropicale* 8: 69–94.

Brajnikov, B. 1953. Les pains-de-sucre du Brésil: sont-ils-enracinés? *Comptes Rendus Sommaire et Bulletin de la Société Géologique de France* 6: 267–269.

Brook, G.A. 1978. A new approach to the study of inselberg landscapes. *Zeitschrift für Geomorphologie* Supplement-Band 31: 138–160.

Campbell, E.M. and Twidale, C.R. 1991. The evolution of bornhardts in silicic volcanic rocks, Gawler Ranges, South Australia. *Australian Journal of Earth Sciences* 38: 79–93.

Dale, T.N. 1923. The commercial granites of New England. United States Geological Survey Bulletin 738.

Dutton, C.E. 1882. The physical geology of the Grand Cañon district. In Second Annual Report of the United States Geological Survey 1880–81. Government Printing Office, Washington. pp. 49–166.

Falconer, J.D. 1911. The Geology and Geography of Northern Nigeria. MacMillan, London.

Fisher, O. 1866. On the disintegration of a chalk cliff. *Geological Magazine* 3: 354–356.

Fisher, O. 1872. On cirques and taluses. *Geological Magazine* 8: 10–12.

Godard, A., Lagasquie, J.-J. and Lageat, Y. 1994. Les Régions de Socle. Faculté de Lettres et Sciences Humaines de L'Université Blaise Pascal. Paris.

Herrmann, L.A. 1957. Geology of the Stone Mountain-Lithonia district, Georgia. Georgia Geological Survey Bulletin 61.

Hibbard, G.R. 1994. The tragedy of Hamlet, Prince of Denmark of William Shakespeare. Oxford University Press. Oxford U.K.

Hill, S.M., Ollier, C.D. and Joyce, E.B. 1995. Mesozoic deep weathering and erosion: an example from Wilsons Promontory, Australia. *Zeitschrift für Geomorphologie* 39: 331–339.

Holmes, A. 1918. The Pre-Cambrian and associated rocks of the District of Mozambique. *Quarterly Journal of the Geological Society of London* 74: 31–97.

Holmes, A. and Wray, D.A. 1912. Outlines of the geology of Mozambique. *Geological Magazine* 9: 412–417.

Ikeda, H. 1991. The inselberg topography of northern Kenya. *Memoirs of Nara University* 19: 79–98.

Jeje, L.K. 1973. Inselbergs' evolution in a humid tropical environment: the example of south western Nigeria. *Zeitschrift für Geomorphologie* 17: 194–225.

Jessen, O. 1936. Reisen und Forschungen in Angola. Reimer, Berlin.

Jutson, J.T. 1914. An outline of the physiographical geology (physiography) of Western Australia. Geological Survey of Western Australia Bulletin 61.

Keyes, C.R. 1912. Deflative scheme of the geologic cycle in an arid climate. *Geological Society of America Bulletin* 23: 537–562.

King, L.C. 1949. A theory of bornhardts. *Geographical Journal* 112: 83–87.

King, L.C. 1968. The origin of bornhardts. *Zeitschrift für Geomorphologie* 10: 97–98.

Kleman, J. 1994. Preservation of landforms under ice sheets and ice caps. *Geomorphology* 9: 19–32.

Lamego, A.R. 1938. Escarpas do Rio de Janeiro. Departamento Nacional da Produção Mineral (Brasil). Servicio Geologia y Mineria Boletin 93.

Le Conte, J.N. 1873. On some of the ancient glaciers of the Sierras. *American Journal of Science and Arts* 5: 325–342.
Lester, J.G. 1938. Geology of the region round Stone Mountain, Georgia. *University of Colorado Studies* (Series A) 26: 89–91.
Lister, L.A. 1987. The erosion surfaces of Zimbabwe. Zimbabwe Geological Survey Bulletin 90.
Mennell, F.P. 1904. Some aspects of the Matopos. 1. Geological and physical features. *Proceedings Rhodesian Science Association* 4: 72–76.
Merrill, G.P. 1897. Treatise on Rocks, Weathering and Soils. Macmillan, New York.
Migon, P. 1993. Kopulowe wzgorza graniowe w kotlinie Jeleniogorskiej. *Czasapismo Geograficzne* 64: 3–23.
Mitra, S. 1992. Balanced structural interpretations in fold and thrust belts. In Mitra S. and Fisher G.W. (Eds.) Structural Geology of Fold and Thrust Belts. John Hopkins University Press, Baltimore. pp. 53–77.
Mountford, C.P. 1965. Ayers Rock. Its People, their Beliefs and their Art. Angus and Robertson, Sydney.
Passarge, S. 1895. Adamaua. Bericht uber die Expedition des Deutschen Kameroun-Komitees in den Jahren 1893–94. Reimer, Berlin.
Post, L. van der 1958. The Lost World of the Kalahari. Penguin, Harmondsworth.
Rodríguez, R. 1994. Control estructural y morfología granítica asociada. Los "Penedos de Traba" (NW, Galicia, España). Geomorfología en España, S.E.G.,Logroño, 73–83.
Ritchot, G. 1975. Essais de Geomorphologie Structurale. Les Presses de l'Université Laval, Quebec.
Rognon, P. 1967. Le Massif de l'Atakor et ses bordures (Sahara Central). Etude Géomorphologique. Centre National de Recherche Scientifique, Paris.
Toit, A.L. du 1937. Geology of South Africa. Oliver and Boyd, Edinburgh.
Twidale, C.R. 1978. On the origin of Ayers Rock, central Australia. *Zeitschrift für Geomorphologie* Supplement-Band 31: 177–206.
Twidale, C.R. 1980. The origin of bornhardts. *Journal of the Geological Society of Australia* 27: 195–208.
Twidale, C.R. 1982. Les inselbergs à gradins et leur signification: l'exemple de l'Australie. *Annales de Géographie* 91: 657–678.
Twidale, C.R. 1986. Granite platforms and low domes: newly exposed compartments or degraded remnants? *Geografiska Annaler* (Series A) 68: 399–411.
Twidale, C.R. and Bourne, J.A. 1975. Episodic exposure of inselbergs. *Geological Society of America Bulletin* 86: 1473–1481.
Twidale, C.R. and Bourne, J.A. 1978. Bornhardts. *Zeitschrift für Geomorphologie* Supplement-Band 31: 111–137.
Vidal Romaní, J.R. 1989. Granite geomorphology in Galicia. *Cuadernos do Laboratorio Xeolóxico de Laxe* 13: 89–163.
Vidal Romaní, J.R. 1991. Kinds of fabric and their relation to the generation of granite forms. *Cuadernos do Laboratorio Xeolóxico de Laxe* 16: 301–312.
Vidal Romaní, J.R. and Twidale, C.R. 1998. Formas y Paisajes Graníticos. Serie Monografias 55, Universidade da Coruña, Servicio de Publicacións.
Watts, W.W. 1903. Charnwood Forest, a buried Triassic landscape. *Geographical Journal* 21: 623–633.
Whitlow, J.R. 1978–9. Bornhardt terrain on granitic rocks in Zimbabwe: a preliminary assessment. *Zambian Geographical Journal* 33–34: 75–93.
Willis, B. 1934. Inselbergs. Association of American *Geographers Annals* 24: 123–129.
Willis, B. 1936. East African plateaus and rift valleys. In Studies in Comparative Seismology. Carnegie Institute, Washington DC, Publication 470.

7

Other granitic residuals and uplands

Though bornhardts are a common, and, it is argued, basic granitic form, other residuals are also well developed. Some are small, but others take the form of massifs, of which, however, bornhardts and other forms are the basic component, and some of which, viewed regionally, are themselves inselbergs.

7.1 ISOLATED RESIDUALS

7.1.1 *Nubbins*

Nubbins are hills strewn with blocks or boulders (Figs 1.2c and 7.1). Overall, they are roughly domical, though some are bevelled. Nubbins are particularly common and well-developed in the tropical monsoon lands, especially South East Asia, where they are the characteristic granite upland. They are also found in monsoon lands with a distinct dry season, as in northern Australia, and in areas which experience warm humid seasons such as the Serra da Estrela of central Portugal. Where nubbin landscapes are found outside the humid tropics, they denote either inheritance from a former humid climate as in the Mojave Desert (Oberlander, 1972) and other parts of southern California, or local wet sites, as for instance, in the intermontane ranges near Alice Springs, central Australia, and in the Llano of central Texas. They are developed in valley floors in the Swakoprivier valley, south of Karibib in central Namibia (Fig. 7.2), and the White River valley of Mpumalanga.

Where nubbins and bornhardts are found side-by-side, as, for instance, in the western Pilbara, Western Australia, the two types are found to be constructed of apparently the same bedrock, so

Figure 7.1.   Granite nubbin in the western Pilbara of Western Australia.

154 *Landforms and Geology of Granite Terrains*

Figure 7.2. (a) Sketch of the Swakop valley in central Namibia, showing nubbins in valley floor, domes at higher levels, and palaeosurface (X). (b) Nubbins developed in wet valley-floor site in Namaqualand (Western Cape Province), South Africa.

Figure 7.3. Sheet fracture exposed in gneissic nubbin, western Pilbara, Western Australia.

that the contrast between the dome and nubbins is not due to petrological characteristics. Moreover, both sheet fractures and orthogonal fractures are present.

At several sites sheet structure is preserved and domical forms can be seen beneath the outer cover of blocks and boulders (Fig. 7.3) suggesting that nubbins are domes, the outer shells of

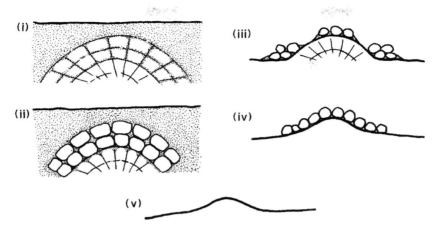

Figure 7.4. Suggested development of nubbins.

which have been broken down to blocks and boulders through the exploitation of fractures disposed normal to the domical surface and which may be related to tensional stresses in the crests of antiforms (Ritchot, 1975).

There is considerable evidence that nubbins are associated with palaeosurfaces. The flattish crests of nubbins in northwest Queenland, and near Alice Springs, Northern Territory, are readily correlated with Late Tertiary palaeosurfaces of which they are thought to be part, possibly in etch form, and below which deep differential compartment weathering took place under warm, humid conditions. Handley (1952) has shown that what he called the tors (which are in reality nubbins) of Tanganyika evolved beneath the African land surface of Early Cainozoic age. Those of the western Pilbara developed beneath either the Cretaceous Hamersley Surface, or the lower Eocene surface on which the Robe River pisolite was deposited.

Thus, it is suggested that most nubbins have their origin in the subsurface and evolve particularly well in warm, humid climates, where weathering is sufficiently aggressive to cause the blocky disintegration of the outer shell or shells of the convex-upward masses of still fresh rock (Fig. 7.4). After the lowering of the plains and the exposure of the residuals, the continued breakdown of the blocks and boulders, and particularly the evacuation of the interstitial grus, causes them to become disarranged as they tumble downslope under gravity. In this way, parts of the inner dome are revealed.

7.1.2. *Castle koppies*

Castle koppies are comparatively small, steep-sided castellated residuals (Figs 1.2d and 7.5). They stand in isolation, and Godard (1977) perceptively refers to castle koppies as inselbergs de poche, indicating that the castellated forms are small compared with bornhardts. In any given area they are lower and really less extensive than their domical counterparts. Orthogonal fractures dominate koppies, and like bornhardts, some display evidence, in the form of zones of flares well above the present plain level, of phased development (Fig. 7.6).

Domed and castellated inselbergs developed in similar rock types coexist in the Harare-Mrewa-Marandellas area of Zimbabwe and elsewhere. Again, at many sites castellated forms rest on large-radius domes (Fig. 7.7) formed on identical bedrock. The angular morphology of castle koppies undoubtedly reflects either widely spaced orthogonal fractures or a well-developed vertical or near-vertical foliation. Thus, on Dartmoor, southwestern England (Simmons, 1964; Worth, 1953; Waters, 1964) the Massif Central of France, the Bohemian Massif in the Czech Republic (Demek, 1964), the Karkonosze Mountains of southern Poland (Jahn, 1962), the Sierra de Gredos, central Spain, the Traba Mountains (Costa da Morte, Galicia), the Pitões dos Junhas in the Sierra de Xurès

Figure 7.5. Castle koppies: (a) Haytor, Dartmoor, southwestern England; (b) in the Andorran Pyrenees.

of southern Galicia NW, Spain (Vidal Romaní, 1989) and northern Portugal (Vidal Romaní, Brum, Zézere, Rodrigues et al., 1990) and in many other places, castellated granite inselbergs are associated with orthogonal joint systems. But not all granites with well-developed orthogonal joint sets give rise to castle koppies. In parts of the Pyrenees (Vidal Romaní, Vilaplana, Martí and Serrat, 1983), though koppies occur on summit plains and crests in Andorra for example, they are absent or scarce elsewhere, even in what appears to be suitably structured rocks. If castle koppies were consistently developed where orthogonal fractures are prominent, they would be widely distributed, whereas they are, perhaps, the least common of the three inselberg forms discussed here.

Many of the koppies of Zimbabwe are developed on gneisses with a well-developed, steeply-dipping, widely-spaced foliation. Gneissic koppies also occur in the eastern Mt Lofty Ranges, near Adelaide, South Australia, and in the southwest of Western Australia, e.g. in Castle Rock, in the

Other granitic residuals and uplands 157

Figure 7.5. (c) In the central Sahara (Rognon, 1967).

Figure 7.6. Castle Rock, in the Mt Manypeaks area, south coast of Western Australia; note flared sidewalls (X).

Mt Manypeaks district, east of Albany (Twidale, 1981 see Fig. 7.6). But well-foliated gneiss is not, in itself, a guarantee of castellated form and, on the other hand, many bornhardts are developed on gneissic rocks and many castellated hills are formed in granite. Though structure is a significant factor in the genesis of castle koppies, it is not an overriding one: evidently the structural base has been exploited in particular ways and conditions, in order for the angular form to evolve.

Some workers associate koppies with particular climates and processes. Godard (1997) considers that castellated inselbergs are either due to frost shattering, or are large residual masses remaining after differential subsurface weathering, or are the cores of inselbergs remaining after either scarp

158   *Landforms and Geology of Granite Terrains*

Figure 7.7. Castellated forms standing on the crests of granite domes: (a) Remarkable Rocks, southwest coast of Kangaroo Island, South Australia; (b) Devil's Marbles, central Australia.

retreat or differential rock disintegration. Linton (1955, 1964) attributed the Dartmoor tors or koppies to differential compartment weathering under humid, tropical conditions and subsequent exposure of the upstanding residuals by solifluxion and cold-climate processes generally. According to Demek (1964) the koppies of Bohemia are Fernlinge remaining after scarp retreat, but he reached conclusions similar to those of Linton (1964) regarding the processes and mechanisms involved.

Cold conditions have also been invoked in explanation of koppies on the Iberian Peninsula. Frost action could have further shattered and mobilised the earlier developed regolith, to cause further steepening of the masses of fresh rock protruding into the regolith. Alternatively, freeze-thaw

Other granitic residuals and uplands 159

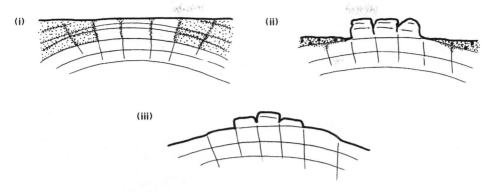

Figure 7.8. Suggested development of castle koppies.

action in the scarp-foot zone during the gradual exposure of the residuals could produce the cliffed bounding slopes. In the central Serra de Gêrez, in the northwestern Iberian Peninsula, the flanks of the castle koppies or borrageiros, which in glacial times were small nunataks, have been steepened as a result of the preglacial regolith having been bulldozed away by local glaciers (Vidal Romaní, Brum, Zézere, Rodrigues et al., 1990). Minor etch forms such as rock basins were eliminated at the same time though they survive in areas bordering the glaciers, where the regolith remains essentially intact. Glaciers also evacuated scree debris accumulated during interglacial nival episodes. Both of these mechanisms contributed to the exposure and shaping of the koppies in coherent rock. Similar chronologies can be reconstructed to account for the castellated landscapes of many of the Hercynian uplands of western Europe, for instance in such regions as the Massif Central, as well as Alpine fold belts like the Pyrenees.

On the other hand, cold climate processes can hardly be invoked in explanation of the well-known koppies of Zimbabwe (Fig. 1.2d), or those of mid-latitude deserts such as the Sahara (Fig. 7.5c). The location of some koppies on the crests of large-radius domes (Fig. 7.7) suggests that they are the last remnants of massive sheet structures, the marginal areas of which have been worn away. Only the crestal zones are preserved, suggesting that there has been strong marginal attack, most likely by subsurface moisture (Fig. 7.8). Several conditions are especially conducive to pronounced marginal attack. In arid and semi-arid lands there is a tendency to steep slopes due to the contrast between active weathering in the moist subsurface and the stability of the dry, and hence stable, exposed surfaces. Whereas nubbins are initiated wholly in the subsurface, koppies may have evolved where the crests of dome structures are exposed and therefore relatively stable. Local wet sites are also conducive to intense marginal weathering, as, for instance, at the Devil's Marbles, Northern Territory (Fig. 7.9). Long periods of land-scape stability such as are associated with the various land surfaces in such interior sites as Zimbabwe, a country rich in koppies, allow even modestly aggressive weathering processes to have marked effects. All that is necessary is that the upper part of the residual be either exposed or located in the drier, near-surface zones of the regolith, while lower parts are affected by moisture in the deeper regolith. Also, vertical or near-vertical foliation and steeply inclined fractures not only allow moisture to penetrate into the rock, but also impose a measure of structural control on the form of the weathering front.

7.1.3 *Large conical forms or medas*

The medas of the granitic uplands of western Iberian Peninsula are conical in shape (Fig. 7.10a), and similar, if isolated, forms are found elsewhere (Fig. 7.10b). Other, miniature forms (up to 4 m high) are developed in granite at Houlderoo Rocks, in the southern piedmont of the Gawler Ranges, South Australia (Twidale, Mueller and Campbell, 1988 and see Fig. 4.8). The latter have evolved in a wet site, the piedmont zone of a major massif, where there is evidence of subsurface weathering to the depth of a few metres in the Late Cainozoic. For example, there are flared basal

160 *Landforms and Geology of Granite Terrains*

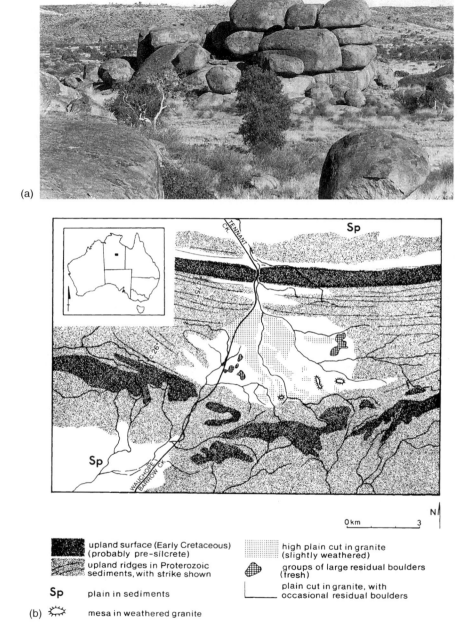

Figure 7.9. (a) Castle koppie at the Devil's Marbles, Northern Territory. (b) Map of locality. Note location in valley bordered by bevelled quartzite ridges.

slopes (see Chapter 8) up to 5 m high in some of the valleys, and a platform standing 2–3 m above the present scarp foot bordering the piedmont plains. The conical hills are, it is suggested, due to intense weathering of fracture-defined blocks, the crests of which were exposed. The crests were thus protected, but below ground weathering proceeded apace, especially in the near surface zone

Other granitic residuals and uplands 161

Figure 7.10. (a) A conical residual, the Meda de Rocalva, Serra do Gêrez, northern Portugal-southern Galicia. (b) A meda, known as the Sugarloaf, in the Organ Mountains, southern New Mexico (Seager, 1981).

where water table fluctuations caused repeated cycles of wetting and drying, and where organic inputs were optimal. The rock just below the surface was rapidly weathered and the rate of weathering diminished gradually in depth. Shallow groundwaters evidently did not stand and cause intense weathering a metre or two below the surface for flared forms are subdued. The surface exposure was reduced to a small area. The effectiveness of weathering evidently diminished gradually with depth beneath the surface to produce inclined, rectilinear slopes, and thus a conical form overall (Fig. 7.11).

Whether the larger medas are of similar origin is not known. The medas of Galicia, NW Spain and northern Portugal are developed in a fine-grained leucogranite with well-developed steeply

Figure 7.11. (a) Suggested evolution of conical residuals. (b) Tower tending to acute cone-shape in granite surrounded by periglacial scree from Pitões das Junhas, Serra do Gerês, northern Portugal/southern Galicia, Spain.

inclined fracture sets. The borrageiros or koppies, with which they are spatially associated, are found in coarse-grained granites. Both conical and castellated forms are of two-stage origin.

7.1.4 Towers and acuminate forms

Towers due to the exploitation, most commonly by frost action, of prominent steeply inclined fractures are found in the eastern Sierra Nevada, California (Bateman and Wahrhaftig, 1966), around Mt Whitney, in southern Greenland, in the Tassili of the central Sahara, in the Organ Mountains of southern New Mexico, in Os Pitões das Junhas, northern Portugal/southern Galicia, Spain, in the Patagonian Andes (with the Cerro Torre prominent) in Argentina and in Sabah, East Malaysia (Fig. 7.12). The Organ Mountains stand in a midlatitude desert landscape, but are high enough to attract snow and ice in winter. The Tassili examples stand at high elevations and may have been cold enough in glacial periods of the Pleistocene to induce nival action. Though close to the Equator, northeastern Sabah, standing more than 4000 m above sealevel was sufficiently cold in glacial times of the later Cainozoic for nival processes to produce prominent acuminate peaks (Fig. 7.12c).

Elsewhere, towers of rock rather than domes or koppies protrude above the general level of the massif or slope. Cathedral Rocks, in the Yosemite of California, are a good example (Huber, 1987 and see Fig. 7.13). Again, well-developed vertical or near vertical fractures have been exploited.

Other granitic residuals and uplands 163

Figure 7.12. Acicular towers in (a) southern Greenland (Oen Ing Soen); (b) the Organ Mountains, New Mexico.

7.2 MASSIFS

Granite uplands are found in many parts of the world. Their morphology varies according to elevation, degree of dissection, and climate. Much of the Labrador Plateau in Canada, for example, comprises a high granitic plain characterised by low bosses and rounded hills (Fig. 7.14) which stand in marked contrast with the Labrador Trough, a region of folded sediments and basic volcanics on which is developed ridge and valley topography. The region was covered by ice sheets until a few thousand years ago. The ice sheets have left a legacy of a scoured etch plain, moraines and basins of interior drainage, so that large areas are occupied by lakes and swamps; it is for good reason

164 *Landforms and Geology of Granite Terrains*

Figure 7.12. (c) St John's Peak, Sabah, East Malaysia (Photobank, Singapore).

Figure 7.13. Cathedral Rocks, Sierra Nevada, Yosemite, California.

Other granitic residuals and uplands 165

Figure 7.14. General view of Labrador Plateau, north of Schefferville, Province of Québec, Canada, showing scoured granite surface with rounded, part-buried, boulders.

Figure 7.15. Serra do Gerês, northern Portugal/southern Galicia, Spain.

that the region is known as the lake plateau. Nival processes have resulted in such features as altiplanation benches and patterned ground. But the granitic character of the bedrock finds expression in unspectacular yet typical granite forms.

The Serra do Gerês, in southern Galicia/northern Portugal, is a massif of Hercynian granites. Differential weathering of fracture zones has been exploited by Late Pleistocene glaciers, as well as by Holocene frost and river action. All of the major morphological features are two-stage features initiated by subsurface weathering but exposed and brought into relief by glaciers, nival activity and running water. The straight valleys of the massif reflect major fracture zones, and the medas or conical residuals also stand in rows in zones of compression (Fig. 7.10a). Castellated forms or borrageiros occur in zones of greater fracture density (Fig. 7.15). Roches moutonnées are

Figure 7.16. (a) The Beardown Man – a small menhir. (b) Pseudobedding at Wattern Tor, Dartmoor, southwestern England.

relic from a Late Pleistocene glacial phase and solifluxion deposits, screes and wedging from recent and contemporary frost action.

The Dartmoor massif of southwestern England is a high plain, the borders of which are deeply dissected by such rivers as the Okement, Tavy, Teign, Tamar and Dart which together form a radial pattern (Waters, 1964 and see Fig. 1.12a). In the core of the massif, which over wide areas rises to over 450 m above sealevel, there are large expanses of high boggy plain surmounted by low but prominent tors, or castle koppies (Fig. 7.5), and with small menhirs such as the Beardown Man (which were signposts or landmarks in former times – Fig. 7.16a), as well as stone circles and other ancient monuments. Outcrops are characterised by pseudobedding which cuts across sheet

Other granitic residuals and uplands 167

Figure 7.17. Koppie in the Margeride, central France.

fractures (Figs 2.2a and c, and 7.16b). Rock basins are well-developed (Fig. 1.2f), as are frost shattered plates, known locally as clitter. Details vary but most workers consider that the tors were initiated by differential, structurally-controlled subsurface weathering during the Early Tertiary, in warm humid conditions, and that the weathered mantle was later stripped, in part at least by solifluxion during Pleistocene glacial phases. The Massif Central, and the Margeride, also in central France, are morphologically similar (Fig. 7.17).

The granitic (monzonite) areas of the Sierra Nevada of California (Bateman and Wahrhaftig, 1966) are characterised by deeply dissected (relief amplitude circa 600–650 m) glaciated river valleys such as that of the Merced River. The intervening high plain or plains, such as the Dana Plateau, are remnants of ancient, largely unglaciated planation surfaces which have been stripped of their regolith and which are thus of etch type (Fig. 1.1g). Superbly developed and exposed domes or bornhardts are revealed in and adjacent to the dissected areas. Sheet structure is well-developed, including banks of thin sheets on cirque headwalls and other recently sculpted surfaces (Fig. 2.7c). Rock basins and slots are also prominent.

In the humid tropics, with high temperatures, abundant moisture and high inputs of organic chemicals and activities, weathering is rapid, with the alteration of micas measured in decades and of feldspars in centuries (Caillère and Henin, 1950). Silica, especially in an amorphous state, is also dissolved. As a result, the landscape is typically blanketed by a thick regolith, and bedrock is largely obscured. Only on the coast, in river channels, on steep slopes, and where there has been vegetation clearance for agriculture or as a result of construction works (road cuttings, quarries, tunnels and adits), the fresh bedrock is exposed. The overall topography is one of all slopes (Fig. 7.18a), though frequently with numerous blocks and boulders (nubbins) and with a tendency to convex slopes and hence domical morphology. River patterns are frequently angular and are in many instances demonstrably related to fracture patterns in the local bedrock.

Though limited, such exposures as suggested in other environments, weathering is characteristically incomplete, with corestones or core-boulders, to use Scrivenor's (1931) apt term, widely preserved (Fig. 5.4). Mass movements of debris are common and as a result of these, plus stream erosion, many such corestones are exposed as boulders. The evacuation of detritus has led to boulders and related forms, such as miniature towers, being exposed on peaks and ridge crests. The steep, bare and frequently slightly convex-outward rock faces found on some steep hillsides are construed as sheeting planes.

Figure 7.18. (a) Dissected granitic terrain in Sabah, East Malaysia (Photobank, Singapore). (b) Boulder field.

The all slopes topography typical of many parts of the humid tropics can be interpreted as due to fracture-controlled stream incision, with the rate of downcutting controlled by local and regional baselevels, but almost everywhere less than the rate of weathering. As noted (Chapter 6), most bornhardts are two-stage or etch forms, and nubbins are bornhardts the outer sheets of which have been broken down, mostly below the land surface, to blocks and boulders. They are typical of monsoon areas and reflect the rapidity of weathering in such areas. The granitic ridges and ranges of the humid tropics can be regarded as elongate bornhardt complexes which have been converted to nubbins. However, despite stream incision, and reflecting the typical high rate of weathering, they still carry a regolithic cover. They are two-stage forms, the development of which has been arrested at stage one.

Of the minor forms found on granite outcrops (see Figs 7.18b and c and Chapters 8–11 inclusive), Rillen are common both on the coast and inland, especially in cleared areas. Rock basins and pitting are well-developed and split rocks are known. Neotectonic forms are also notably absent though this may be more apparent than real. Such forms as A-tents, or pop-ups, may be developed at the weathering front, at the base of the regolith, but not be exposed save in special and rare circumstances.

Other granitic residuals and uplands 169

Figure 7.18. (c) Flared boulder, both in the Tampin district of West Malaysia.

Figure 7.19. All slopes topography in granite (a) southwest of Rio de Janeiro, southeastern Brazil.

Thus, in summary, the granite terrains of the humid tropics can be construed as nubbin-dominated uplands, with many characteristic minor weathering and possibly also neotectonic forms developed, but still covered by the regolith except where for one reason or another the latter has been partly or entirely stripped.

7.3 REGIONS OF ALL SLOPES TOPOGRAPHY

All slopes topography in granite has been reported from regions as scattered and climatically diverse as the Sinai Peninsula and the Red Sea Hills of Egypt (Hume, 1925), the hinterland of Rio de Janeiro in southeastern Brazil, the arid northern Flinders Ranges of South Australia, the Peruvian Andes, the uplands of Papua New Guinea and the uplands of southern Poland (Fig. 7.19a and b).

170 *Landforms and Geology of Granite Terrains*

Figure 7.19. (b) Northern Flinders Ranges, South Australia (Publicity and Tourist Bureau, South Australia).

Figure 7.20.   Faceted slopes in granite in northwest Queensland (CSIRO).

The topography appears to evolve by stream dissection of massifs in which sheet structure is poorly developed or absent, or in regions in which weathering has eliminated, or all but destroyed, structural influence of the country rock. The landform assemblage can be explained by considering the sequence of events following the stream dissection of a duricrusted surface. The sidewalls of the valley consist of faceted slopes (Fig. 7.20); the bluff in some areas being coincident with a duricrust capping (Figs 1.1h, 4.12 and 7.20). Examples of faceted granitic slopes associated with duricrusts are commonplace, but the duricrust is not essential to their formation. Such faceted slopes are maintained for some time, but because of the basal sapping of the bluff the latter is gradually reduced from below, until it is eliminated. The slopes then consist of essentially rectilinear debris facets (bedrock but with a veneer of detritus). They intersect in sharp crests or aretes: all slopes, for there are no areas of subdued relief, apart from minor flood plains in the valley floors (Fig. 7.21). It is not suggested that duricrusts are essential to the development of all slopes for as explained in Chapter 6, dry upper zones of granite are relatively stable and have similar effects.

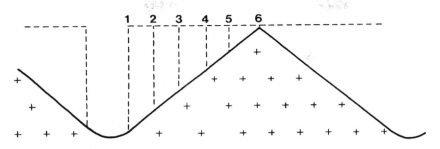

Figure 7.21. Suggested mode of development of all slopes topography.

7.4 DISCUSSION

Bornhardts, nubbins and castle koppies are genetically related forms, the two last-named being derived from the marginal subsurface weathering of domed forms. Nubbins and castle koppies are the reduced remnants of bornhardts. Elements of the two forms are developed in the same hill at some sites. Thus, in the southern face of Blackingstone Rock, on eastern Dartmoor, southwestern England, huge quadrangular blocks are exposed, and the form is castellated, but the northern side is dominated by massive convex-upward slabs of rock that give a domical shape to the hill (Simmons, 1964 and see Fig. 2.12). According to Holmes (1918), Mt Kobe, in Mozambique, also displays contrasted morphology on opposed aspects, and some of the residuals at the Devil's Marbles, Northern Territory, embrace elements of nubbins and koppies (cf. Figs 7.7b and 7.9).

The basic form, the bornhardt, is a structural feature developed on masses that are compact by virtue of their being in stress. They are characterised by, and owe their domical shape to, the development of sheet structure as a result of shortening. Nubbins and castle koppies, on the other hand, though strongly influenced by structure and also found in multicycle landscapes, are in some measure morphogenetic features. Nubbins are best developed in humid, tropical areas as a result of the superficial disintegration of the outer shells (sheet structure). Castle koppies, on the other hand, are due to lateral or marginal weathering, also in the subsurface, and under various climatic conditions (humid, tropical, arid or semi-arid, arctic or subarctic).

Weathering eventually reduces both nubbins and koppies, though the latter especially are very durable. What remain are small domes which are frequently scarcely more than low, convex-upward platforms with a scatter of blocks and boulders. They are nevertheless genetically related to bornhardts. The dome structure is the starting point of an evolutionary sequence that can follow varied paths. All three prominent inselberg forms are initiated in the subsurface.

REFERENCES

Bateman, P.C. and Wahrhaftig, C. 1966. Geology of the Sierra Nevada. In Bailey E.H. (Ed.), Geology of Northern California. California Division of Mines and Geology Bulletin 190. pp. 107–172.
Caillère, S. and Henin, S. 1950. Etude de quelques altérations de la phlogopite à Madagascar. *Comptes Rendues des Séances. Academie des Sciences de Paris* 230: 1383–1384.
Demek, J. 1964. Castle koppies and tors in the Bohemian Highland (Czeckoslovakia). *Biulytyn Peryglacjalny* 14: 195–216.
Godard, A. 1977. Pays et Paysages du Granite. Presses Universitaires de France, Paris.
Handley, J.R.F. 1952. The geomorphology of the Nzega area of Tanganyka with special reference to the formation of granite tors. C.R. Congr. géol. International (Algiers). 21: 201–210.
Holmes, A. 1918. The Pre-Cambrian and associated rocks of the District of Mozambique. *Quarterly Journal of the Geological Society of London* 74: 31–97.
Huber, N.K. 1987. The geologic story of Yosemite National Park. United States Geological Survey Bulletin 1595.
Hume, W.F. 1925. The Geology of Egypt. Volume I. Government Printer, Cairo.

Jahn, A. 1962. Geneza skalek granitowych. *Czasofismo Geograficzne* 33: 19–44.
Linton, D.L. 1955. The problem of tors. *Geographical Journal* 121: 470–487.
Linton, D.L. 1964. The origin of the Pennine tors – an essay in analysis. *Zeitschrift für Geomorphologie* 8: 5–24.
Oberlander, T. 1972. Morphogenesis of granitic boulder slopes in the Mojave Desert, California. *Journal of Geology* 80: 1–20.
Ritchot, G. 1975. Essais de Géomorphologie Structurale. Presses Universitaires Laval, Québec.
Simmons, I.G. (Ed.), 1964. Dartmoor Essays. Devonshire Association for the Advancement of Science, Literature and Art, Torquay.
Scrivenor, J.B. 1931. Geology of Malaya. Macmillan, London.
Seager, W.R. 1981. Geology of Organ Mountains and southern San Andres Mountains, New Mexico. New Mexico Bureau of Mines and Mineral Resources Memoir 36.
Twidale, C.R. 1981. Granite inselbergs: domed, block-strewn and castellated. *Geographical Journal* 147: 54–71.
Twidale, C.R., Mueller, J.E. and Campbell, E.M. 1988. Formes nettoyées dans les roches volcaniques acides en Australie du Sud et au Nouveau Mexique (Etats-Unis). *Revue de Géomorphologie Dynamique* 37: 113–126.
Vidal Romaní, J.R. 1989. Granite Geomorphology in Galicia, N.W. Spain. *Cuadernos Laboratorio Xeoloxico de Laxe* 13: 89–163.
Vidal Romaní, J.R., Brum, A. de, Zézere, J., Rodrigues, L.M. and Monge, C. 1990. Evolución cuaternaria del relieve granítico en la Serra de Gerez-Xures, (Minho Portugal norte y Ourense, Galicia). *Cuaternario y Geomorfología* 4: 3–12.
Vidal Romaní, J.R., Vilaplana, J.M., Martí, C. and Serrat, D. 1983. Rasgos del micromodelado periglacial actual sobre zonas graníticas de los Pirineos españoles (Panticosa, Huesca y Cavallers, Lleida). *Acta Geológica Hispánica* 18: 55–65.
Waters, R.S. 1964. The Pleistocene legacy to the geomorphology of Dartmoor. In Simmons I.G. (Ed.), Dartmoor Essays. Devonshire Association for the Advancement of Science, Literature and Art, Torquay. pp. 73–96.
Worth, R.H. 1953. Worth's Dartmoor. In Spooner G.M. and Russell R.S. (Eds). David and Charles, Newton Abbott.

8

Minor forms developed on steep slopes

Inselbergs are characteristically steep-sided. In large measure, this reflects the high dip of the orthogonal fractures that delineate the forms and of the sheeting joints that plunge steeply near the margins of compartments. In addition, however, the basal slopes of many bornhardts have been oversteepened by weathering and erosion. In particular, flared slopes and related forms are well-developed, and many steep slopes are also grooved or fluted.

8.1 FLARED SLOPES

8.1.1 *Description and characteristics*

Flared slopes are concavities shaped in steeply inclined rock walls. They most commonly occur at the bases or lower margins of hills and boulders, but are also found high on the slopes of bornhardts (see Chapter 6). They are up to 12 m high, though most are less than 4 m (Fig. 8.1a). Some of the concavities are so pronounced that the slopes are overhanging (Fig. 8.2a). In places the concavity is not simple, but consists of two or three minor flares superimposed on the overall concavity (Figs 8.1a

(a)

Figure 8.1. (a) Flared slope, some 14 m high, and known as Wave Rock, on northern flank of Hyden Rock, southern Yilgarn of Western Australia.

174 *Landforms and Geology of Granite Terrains*

Figure 8.1. (b) Flared slope 6–7 m high, southern side of Pildappa Hill, northwestern Eyre Peninsula, South Australia.

Figure 8.2. (a) Overhanging flare some 8 m high developed on point of spur at Ucontitchie Hill. (b) Multiple flares on side of Ucontitchie Hill.

and 8.2b). The steepened slope is not everywhere disposed in horizontal zones, but rather follows the hill-plain or rock-soil junction (Fig. 8.2c). Flared slopes attain their maximum development (that is, greatest degree of overhang and length of steep walls) in two topographic situations. They are well-developed on the points of spurs, as, for example, at Ucontitchie Hill, Eyre Peninsula, South Australia, and also in broad fracture-controlled embayments, for example, Wave Rock, located on the northern side of Hyden Rock, Western Australia (Twidale, 1962 and see Figs 8.1a and 8.3). They

Minor forms developed on steep slopes 175

Figure 8.2. (c) Flares sloping in parallel with soil-rock junction, Ampidianambilahy, Andringitra Massif, Madagascar.

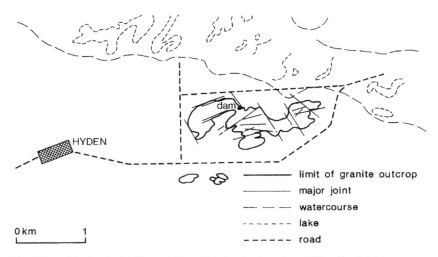

Figure 8.3. Plan of Hyden Rock, Western Australia, showing location of Wave Rock (x).

are found on both the northern and southern sides of residuals, but are particularly well-developed on the latter in southern Australia, though there are exceptions. They are developed on boulders as well as on larger residuals and within clefts as well as marginal to hills and boulders (Fig. 8.4).

They are preferentially exposed on the downslope side of hills and boulders. Flared slopes also occur in subhorizontal zones well above the present plain level, as well as in joint clefts. Excavations and bores on Eyre Peninsula, South Australia, and in Western Australia have shown that they are incipiently developed in the scarp-foot zones of residual hills, beneath the present

176 *Landforms and Geology of Granite Terrains*

Figure 8.4. (a) Flares developed on sidewall of Ucontitchie Hill, northwestern Eyre Peninsula, South Australia, and also on large residual boulder. (b) Boulders with flared sidewalls near Benbarber Corner, Port Kenny, northwestern Eyre Peninsula, South Australia.

plains (Figs 6.20 and 8.5). Moreover (Fig. 8.6), there is some suggestion that the water-table is lower in the scarp-foot zone.

Though flared slopes in granite are apparently not, or only faintly, developed in some areas, even at sites that appear to be suitable, they have been noted in or described from several parts of the world, in cold and warm, arid and humid climates, in interior and coastal areas, lowlands and uplands (Fig. 8.7), and on several rock types, though all are massive (see Chapter 12). They are, however, best and most frequently developed on granitic rocks in shield areas in southern Australia, though they also occur in profusion, well-developed, albeit at a minor scale, in certain

Minor forms developed on steep slopes 177

Figure 8.4.   (c) Flares in cleft Yarwondutta Rock, northwestern Eyre Peninsula, South Australia.

Figure 8.5.   Subsurface flares exposed in excavations (a) at Quarry Hill, near Wudinna, northwestern Eyre Peninsula, South Australia.

other areas such as the Cassia City of Rocks, southern Idaho (Anderson, 1931), in the western USA (Mueller and Twidale, 1988 and see Fig. 8.7c).

8.1.2   *Origin*

Several possible explanations can be suggested to account for flared slopes. The subaerial processes cited as responsible for, or capable of shaping, the flares do not withstand close examination. Wave

178   *Landforms and Geology of Granite Terrains*

Figure 8.5.   (b) In a quarry (X) near Veyrières, southern France; (c) at Kwaterski Rocks, north of Minnipa, northwestern Eyre Peninsula, South Australia.

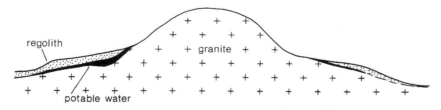

Figure 8.6.   Section through granite hill in southern Western Australia, showing deep water table in piedmont zone (Clarke, 1936).

Minor forms developed on steep slopes 179

Figure 8.7. Boulder with flared sidewalls (a) southern Zimbabwe; (b) Roques Planes Cape, East Catalunya coast, northeastern Spain.

180   *Landforms and Geology of Granite Terrains*

Figure 8.7.   (c) Cassia City of Rocks, Idaho – the Kaiser's Helmet (Mueller and Twidale, 1988). (d) Flared side of cleft on Pearson Islands, Great Australian Bight.

action, considered together with the likely age of the forms, cannot account for their spatial distribution (for example, their occurrence at elevations of some 2,500 m above sealevel in the Sierra Nevada, California, or in Pic Boby, Andringitra Massif, Madagascar (Vidal Romaní, Ramanohison and Rabenandrasana, 1997), and several hundreds of metres above sealevel in the piedmonts of residuals in central Australia (e.g. some 530 m at the base of Ayers Rock). Wind action does not offer a satisfactory explanation for either the local preferred orientation of the steepened slopes or, indeed, for their location in the scarp-foot zone; the latter, being moist, is commonly better vegetated than the surrounding plain and is for that reason protected against any possible sand blast action. Running water fails to account for either the preferred orientation or their development on the points of spurs (where flow diverges). The only explanation that takes account of all the evidence is that the flares are a particular form of etch surface or weathering front developed in the scarp-foot zone as a result of moisture attack on massive rocks and subsequently exposed (Fig. 8.8).

The crucial evidence is seen at such sites as Yarwondutta Rock, Calca Quarry, Quarry Hill, and Chilpuddie Hill, all on northwestern Eyre Peninsula, South Australia, and at Veyrières in southern France, where incipient flared slopes, in the shape of concave sectors of the weathering front, are present beneath the natural land surface (Figs 6.20 and 8.5b), and at various sites on Eyre Peninsula and in the southern Yilgarn of Western Australia where a moat of weathered detritus located around the base of residuals has been demonstrated by augering (Clarke, 1936).

The material with which the concave or flared forms are covered is not transported, introduced detritus but grus, or granite weathered *in situ*. The bedrock surfaces were not formed by epigene

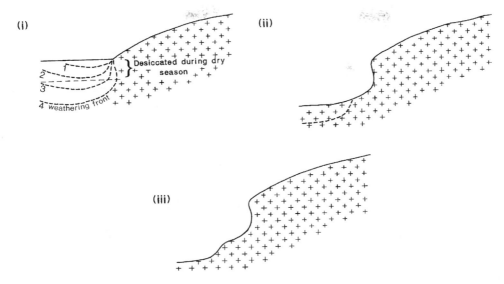

Figure 8.8. Suggested mode of development of flared slopes.

Figure 8.9. Turtle Rock, near Wudinna, Eyre Peninsula, South Australia, showing flares on each side of spur.

processes and then buried. The concavity results from the drying out of the surface and near surface soil and the persistence of moisture and hence longer duration of weathering at depth (Fig. 8.8).

This suggested mode of development explains why incipient forms can be found beneath the regolith, why flared slopes follow the rock-soil junction, and why they tend to be preferentially developed on the shady or wetter sides of inselbergs and in relation to joints which permit the ready infiltration of water into the rock mass. Multiple flares are a consequence either of repetitions of subsurface weathering and lowering of the plain, by surface wash and stream erosion, or water table fluctuations. Volume decrease during weathering, due to evacuation of soluble salts in solution and of fines by flushing, may have resulted in compaction and surface lowering.

Flares are well-developed on the points of spurs because there the fresh rock is attacked from two sides in the subsurface (Figs 8.2a and 8.9). The preferential occurrence of flares on the

182 *Landforms and Geology of Granite Terrains*

(a)

(b)

Figure 8.10. (a) Basal triangular lobes, ribs and scarp-foot depression, Yarwondutta Rock, near Minnipa, Eyre Peninsula, South Australia. (b) Detail of 8.10(a).

(c)

Figure 8.10. (c) Scratches formed by needle-like leaves moved by the wind scraping against algal coating of rock surface, Keith district, South East of South Australia.

downslope side of boulders is due not so much to development as to exposure, for there grus is evacuated downslope, whereas on the upslope side detritus tends to accumulate. Moreover, exceptions are understandable in terms of this working hypothesis. For example, at several sites on northwestern Eyre Peninsula, the major flares occur on the northwestern aspect of the residuals, but in every instance they are developed along major joints; the relative readiness with which water can infiltrate such zones more than compensates for the marked desiccation of the regolith on the exposed northern side of the hills.

Finally, the two-stage, or etch, hypothesis also offers an explanation for the exceptional development of flared slopes in the shield lands of southern Australia, for the region has been comparatively stable for long periods, allowing time for deep weathering and formation of markedly concave, even overhanging weathering fronts and, thus, basal slopes.

On the other hand, incipient flared slopes are not ubiquitously developed in the piedmont zone (Twidale, 1967). There are many excavations where flared slopes, though anticipated, are either absent or poorly developed, at least, not in the vertical range of the exposure. Thus, at Elkington Rock, the upper 5 m of the dome exposed by excavation show only a very faint concavity on its eastern flank (Fig. 6.19a). The flare may be developed at greater depth. Alternatively, however, there may not have been enough runoff from the very small (10 m diameter) platform that was naturally exposed and that is the crest of the dome, or major fractures may be necessary to allow penetration of water sufficient to induce the weathering involved in the formation of a flare.

8.1.3 *Changes after exposure*

After exposure, flared slopes are modified in various ways. Scaling, some possibly inherited from the weathering front (see Chapter 3), causes minor wearing back of the slope. Flutings (see below) extend from near the former ground level and score the concave rock wall. In places separation of flow has caused small plunge pools, and algae are associated with local inversions in the form of channel floors that now stand out as ribs (see below). Triangular divides or lobes, some of them coated with algae, are frequently developed between the lower extremities of flutings (Figs 8.10a and b). They may be caused by weathering by moisture contained in the soil lapping up against the base of the slope and particularly in the small recesses formed by the flutings. In places the algal

184 *Landforms and Geology of Granite Terrains*

coatings accumulated on steep slopes of residual hills and boulders are scratched by branches, twigs and even by needle-like leaves of such plants as spinifex (*Triodia spp.*) moved by the wind and producing arcuate and concentric markings (Fig. 8.10c).

8.2 FRETTED BASAL SLOPES AND OTHER VARIANTS

Though lacking flared slopes as such, many boulders and some inselbergs display basal fretting: the lower sections of the steep bounding slopes of hills and the flanks of blocks and boulders are, to a greater or lesser degree, notched and overhanging (Fig. 8.11). The walls of the concavities are rough, not smooth and regular, as are flared slopes. This, in many instances, reflects the texture of the bedrock, the fretting being especially well-developed in coarsely crystalline or porphyritic granite. Nevertheless, and despite morphological differences in detail, the fretting

Figure 8.11. Basal fretting of (a) residual boulder at Devil's Marbles, Northern Territory; (b) slope of Chilpuddie Hill, northwestern Eyre Peninsula, South Australia.

Minor forms developed on steep slopes 185

appears to be of the same origin as flared slopes, and is a manifestation of soil moisture attack in the scarp-foot zone.

Many examples of basal fretting have been noted in tropical and subtropical, arid and semi-arid regions, where scarp-foot water concentrations are highly significant. But they are by no means restricted to such areas. Thus, on Dartmoor, southwestern England, isolated blocks and menhirs (Fig. 7.16a) display basal fretting and/or basal depressions. The basal fretting around the menhirs argues a rapid rate of development, at least in the geological context.

In some areas, basal indentation occurs at the margins of ephemeral lakes and may be caused by either weak wave attack, or weathering by saline lake waters, or a combination of the two. Jutson (1914), described what he called billiard-table surfaces backed by basally notched cliffs from the Salinaland division of Western Australia, and similar forms have been observed in granite gneiss at the margins of Lake Greenly, on southern Eyre Peninsula, South Australia. But these are

(c)

(d)

Figure 8.11. (c) The Cumberland Stone, from which the Duke of that name directed the Battle of Culloden Moor in 1745 (British Tourist Authority); (d) on large residual boulder with closely-spaced and strongly developed vertical fractures, Cassia City of Rocks, Idaho (Mueller and Twidale, 1988).

186 *Landforms and Geology of Granite Terrains*

Figure 8.12. (a) Acuminate and ensiform residual, Mt Manypeaks area, near Albany, south coast of Western Australia. (b) Hourglass-shaped rock, Murphys Haystacks, west coast of Eyre Peninsula, South Australia.

Figure 8.13. Pedestal or mushroom-shaped rock in granite (a) Sierra Guadarrama, central Spain.

exceptions. Most basal fretting is caused by soil moisture, and the resultant notches, like flared slopes, provide a measure of recent soil erosion.

All-round attack of boulders or columns has produced fastigiate or conical residuals known as Chinaman's, or coolie, hats (see Fig. 8.10d). Acuminate (or pointed) and ensiform (or blade-like) boulders (Fig. 8.12a) are probably of similar origin, as are hourglass forms or dumb-bells (Fig. 8.12b), which vary from the pointed forms only in the degree of the concavity developed. Mushroom or pedestal rocks (Fig. 8.13) are other variants, and there are, in addition, several rocks that are named from their resemblance to specific objects, for example, anvil rocks (Fig. 8.13d) (see also Chapter 9).

Minor forms developed on steep slopes 187

Figure 8.13. (b) Western Sahara desert (Peel, 1941). (c) Sierra Nevada, California (Wahrhaftig, 1965). (d) Anvil-shaped rock, Caloote, western Murray Basin, South Australia.

188 *Landforms and Geology of Granite Terrains*

Figure 8.14. Piedmont angle or nick (a) at base of nubbin at Naraku, northwest Queensland. (b) Base of inselberg in the western Sahara

8.3 SCARP-FOOT WEATHERING AND EROSION, AND THE PIEDMONT ANGLE

One of the contrasts between the arid and semi-arid landscapes of low and middle latitudes, and others, say from the humid mid-latitudes, is that in the former there is an abrupt transition from hill to plain (Fig. 8.14). This sharp break of slope is called the piedmont angle or nick. It has long excited the attention and interest of geologists and geomorphologists and various explanations for it have been advanced. Some have suggested that it marks the junction between adjacent fault blocks, and though the presence of zones of dislocation may assist in and enhance its formation, the piedmont angle is most commonly and well-developed in areas devoid of faults. The feature has been explained as due to lateral undercutting by streams debouching from uplands on to plains,

Figure 8.14. (c) Detail of part of (b) (Peel, 1941). (d) In Namaqualand (Western Cape Province), South Africa.

depositing part of their coarse load and diverting themselves against the mountain front. This suggestion is at odds with the field evidence, for most rivers emerging from uplands continue without pronounced deviation toward the centre of the basin or the axis of the valley. The piedmont angle is developed in sectors lacking surface streams and around residuals too small to spawn streams and rivers. Moreover, there is about the lateral corrasion hypothesis a touch of the chicken and egg argument, for the stream diversion crucial to the explanation is caused by stream deposition consequent on a change in stream confinement and a break of slope; which is what is to be explained. A similar comment applies, but more strongly, to the suggestion that the change of slope reflects a change from turbulent to sheet flow consequent on a change of gradient.

The most reasonable explanation of the piedmont angle is that it is due to pronounced weathering and subsequent erosion. In many sectors this takes place in a scarp-foot zone the location of which reflects structural variations, such as contrasts in fracture density or of lithology. Faults also allow infiltration of water and rapid and deep weathering. But that weathering and erosion of lower slopes alone can cause the piedmont angle is demonstrated in plateau areas where a sharp transition from hillslope to plain is developed in uniform, usually weak, bedrock. For example, in the Tent Hills region, in the arid interior of South Australia, the plateaux and mesas are capped by flat-lying quartzite underlain by shale. It is the latter that is intensely weathered on lower slopes, and in places dissected to give scarp-foot depressions or valleys (see below, Chapter 8). The backing scarp is regraded and steepened to the maximum inclination commensurate with stability, and the piedmont angle is maintained or enhanced.

Flared slopes and other forms of basal fretting are merely a special case of scarp-foot weathering and erosion inducing the development of the piedmont angle. Its especially pronounced development in arid and semi-arid lands in part reflects the increased relative significance of moisture in such climatic regions. It may also be due to the greater likelihood of views of such sharp breaks of slope being uninterrupted by vegetation, for though the moisture responsible for scarp-foot weathering encourages vegetation growth in arid regions, it is localised, whereas in humid areas woods and forests commonly conceal the true nature of the scarp-foot zone.

8.4 ROCK PLATFORMS (SEE ALSO DISCUSSION OF ROCK PEDIMENTS IN CHAPTER 4)

8.4.1 *Description*

Flared slopes extend laterally into gently sloping bedrock surfaces called rock platforms, and for this reason are considered here as well as in Chapter 4. At some sites a depression intervenes between hillslope and platform. Platforms vary in width between a few centimetres at the base of some boulders, and a few metres and several hundreds of metres, as, for instance, at Peella Rock and Corrobinnie Hill in South Australia (Figs 4.6 and 8.15a), and The Humps and Varley Township Hill in Western Australia (Fig. 8.15b), and Ampidianambilahy, Andringitra Massif, Madagascar (Vidal Romaní, Ramanohison and Rabenandrasana, 1997). They vary in extent from a few square metres to several square kilometres. They are most characteristically developed bordering residual hills, but they are also found in isolation, unrelated to any upland mass, though whether this has resulted from the elimination of a former inselberg or whether there never was an upland and the platforms are merely crestal exposures of very large-radius concealed or incipient domes, is a moot point. In detail, some platforms are dimpled, due to the development of shallow, saucer-shaped depressions, and scored by gutters that, together, form a rudimentary drainage network. Small medas, blocks, boulders and patches of grus remain on some platforms.

8.4.2 *Origin*

Platforms are erosional in the sense that they cut across jointing and other rock structures. Like flared bedrock slopes, they have been shown to extend beneath the present regolith and, also like flares, they are regarded as etch forms, or exposed parts of the weathering front. Indeed they are merely lateral extensions of flared slopes. They are especially well-developed in wet sites such as the scarp-foot zone, topographic lows such as the Corrobinnie Depression, South Australia, and bordering ancient watercourses or lakes, for instance, in the southwest of Western Australia. The grus and boulders found on some platforms are remnants of the regolith that formerly covered the planate forms and that has now been partially removed.

8.5 SCARP-FOOT DEPRESSIONS

8.5.1 *Description*

Shallow valleys and moats or linear depressions (Clayton, 1956), also known as Bergfuss-niederungen and dépressions de piedmont, are found around the bases of some inselbergs in arid and semi-arid (especially savanna) landscapes. The valleys are integrated with the local drainage system, but the depressions are enclosed. Such depressions have been reported from West Africa, from the Sudan, where they are known as fules, and from the Egyptian Desert. They occur in central and southern Australia, and in the Mojave Desert of southern California.

Most of the moats are just a few metres across, but some attain widths of several scores of metres, and that around Gebel Harhagit, in Egypt (Dumanowski, 1960), is more than a kilometre wide in places. Some in central Australia, developed around granite nubbins, are of similar dimensions

Minor forms developed on steep slopes 191

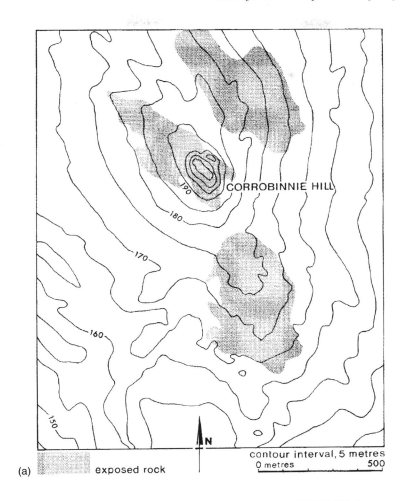

Figure 8.15. (a) Map of Corrobinnie platform, northern Eyre Peninsula, South Australia. (b) Platform at base of granite inselberg at Varley Township, southwest of Western Australia.

192 *Landforms and Geology of Granite Terrains*

(Fig. 8.16). Most are only a metre or two deep, but in others the floor stands several metres below the level of the surrounding plain. At the other end of the scale a small depression developed at Yarwondutta Rock, on Eyre Peninsula, South Australia, is about 7 m long, less than 2 m wide and only 300 cm deep (Figs 8.10a and b).

8.5.2 *Origin*

Dumanowski (1960) favoured a structural (lithological) origin for scarp-foot depressions. Certainly, the specific features he described from the Egyptian Desert are eroded in metamorphic

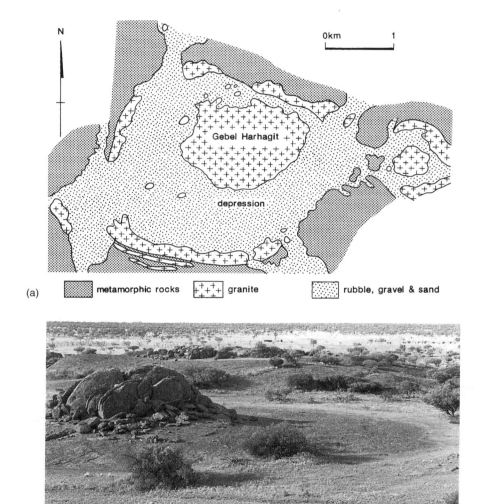

Figure 8.16. Scarp-foot depression (a) at Gebel Harhagit, Egypt, (b) around granite nubbin, central Australia (CSIRO).

rocks, whereas the backing ridge consists of granite. In many other instances, however, no such lithological contrast can be detected. On the contrary, the bedrock in the floor of the depressions is apparently of the same type as that exposed on the adjacent hills. Minor scarp-foot depressions are undoubtedly caused by runoff scouring the inner fringe of the regolith where it laps up against the bare rock hillslope, as, for instance, at Chilpuddie Rock, on northwestern Eyre Peninsula, South Australia. Cotton (1942, p. 42) pointed out that rivers in some arid and semi-arid regions *"consistently hug the mountain bases in a manner that cannot be fortuitous"* and went on to suggest that the linear depressions are river channels. There is here, however, a possible confusion of cause and effect. Moreover, accepting the notion of scarp retreat, Cotton (1942) argued that the mountain fronts and river channel depressions must recede together. This hypothesis is difficult to sustain, for very few of the depressions observed in Australia and North Africa are occupied by streams. In those that are, it may be a case of the drainage channels running in the depressions rather than excavating them. Furthermore, some of the depressions stand at the base of residuals too small to generate streams, and others occur where no streams debouch from the uplands.

The most plausible explanation for scarp-foot depressions is that runoff from the hills saturates the piedmont zone, which is in consequence, intensely weathered. There is ample evidence of such chemical attack in a wide variety of lithological settings. Some products of weathering are evacuated in solution. Some fines may by flushed out, or deflated by the wind. There may be a volume decrease, compaction and settling, and surface subsidence. The water table is lower (Fig. 8.6). Additionally, and as doubtless happens in some places, once formed, the depressions may have been deepened by intermittent or episodic surface streams. Whatever agent is responsible, however, the scarp-foot zone is lowered and a topographic depression aligned along the hill base is formed. The comparatively great breadth of some depressions relative to the associated residual hill, like those featured from central Australia (Fig. 8.16b), suggests that the process of scarp-foot rotting and lowering must be of long duration.

Some regard the forms as having developed in the humid tropics. They believe that where these scarp-foot depressions are found in arid lands they are inherited from former humid climatic phases, but this suggestion is sustained neither by general argument nor by field evidence. Water is, if anything, relatively more important in arid and semi-arid lands than elsewhere and, because of its concentration in the piedmont, achieves results that are more pronounced than in other climatic contexts.

One important result of the scarp-foot weathering and erosion manifested in the several landforms just described is that the piedmont angle, or nick, becomes pronounced. The feature is basically a structural form and in granitic terrains is roughly coincident with the margins of massive compartments. But weathering and surface lowering due to volume decrease, as suggested by Ruxton (1958), or preferential erosion of the zones of altered rock, cause the structural feature to become emphasised. Truly angular nicks are particularly commonplace in granitic rocks where vertical or near-vertical fractures are exploited and exposed by weathering and erosion.

8.6 FLUTINGS OR GROOVES

8.6.1 *Description*

Many steep slopes, bordering boulders as well as hills, and including flared slopes, are scored by grooves, and are said to be fluted (Fig. 8.17 also 1.2g). These grooves are variously known as Silikatkarren, Pseudokarren, Granitkarren, regueros, canales, estrías, zlobki, cannelures, Riefelungen and Kannelierungen, and so on, as well as flutings (Schmidt-Thomé, 1943; Bulow, 1942; Carlé, 1941). Most grooves are U-shaped in cross-section (cf. Rundkarren formed in limestones), others are more open, while many others are broad, shallow and flat-floored. Their width and depth is measured in centimetres (commonly of the order of 20–30 cm) though some a few metres deep have been noted. They are found not only on the exposed outer walls of boulders, blocks and hills, but also on overhanging slopes (Fig. 8.17c) and on the interior walls of tafoni (Wilhelmy,

Figure 8.17. (a) Flutings and Kluftkarren on flanks of granitic bornhardts near Lundi River Bridge, southern Zimbabwe (Lister, 1987). (b) Closely-spaced flutings on inclined sidewall of granite hill near Saldanha, western Cape Province, South Africa. (c) Flared sidewall of inselberg scored by numerous flutings or Rillen, on Pildappa Rock, northwestern Eyre Peninsula, South Australia.

Minor forms developed on steep slopes 195

Figure 8.17. (d) Fluted boulder, Tampin, West Malaysia. (e) Fluted steep sidewall of granite block, Remarkable Rocks, Kangaroo Island, South Australia. Note tafoni in foreground.

Figure 8.17. (f) Fluted overhanging backwall of tafoni at Murphys Haystacks, northwestern Eyre Peninsula, South Australia.

1964 and see Fig. 8.17f; see also Chapter 10). Flutings are well-developed in equatorial areas of consistently high rainfall and were described from Singapore and Malaysia (Hellstrom, 1941) in the middle of the last century and the Andringitra Massif of Madagascar (Petit, 1971, 1990).

The humid tropical provenance of some of the forms has seemingly been confirmed by several workers, but in reality they are found far beyond the confines of such regions. They have been described from such arid and semi-arid areas as central Brazil, and occur also in the arid lands of southwestern Angola, on Domboshawa in Zimbabwe, on Paarl Mountain near Capetown and the Volzberg, in the savannas of Surinam. They are well-developed at several sites in Victoria and on Eyre Peninsula, South Australia. They occur in cool, high altitude areas in the highlands of tropical Brazil. They have been described from recently deglaciated or cold areas such as Scandinavia, and Bohemia, from seasonally cold areas like the Margeride of central France. Flutings occur on granite exposures in many parts of the Iberian Peninsula, coastal and interior, glaciated and nonglaciated. On the other hand, they are strangely lacking in northern Australia; they are not absent, but they are rare, and then poorly developed.

8.6.2 Origin

That a broad and general control is exerted by structure on the development of flutings was first indicated when it was pointed out that the forms are developed only on massive rocks. They do not occur on closely fractured bedrock, save in special circumstances, presumably because the moisture that is most likely responsible for their development can there percolate into the cracks and crevices, and does not attain the volume or velocity required to scour isotropic rock surfaces.

In Swaziland, on the Mswati Granite, some small grooves are preferentially developed on microgranite which has numerous grain boundary cracks susceptible to moisture attack. That some grooves in granite are Kluftkarren (Fig. 8.18), being due to the exploitation of fractures by weathering agencies, was recognised long ago. In 1849 Logan suggested that some of the grooves of Pulau Ubin, in the Johore Strait, may have been developed in linear zones where the granite was less cohesive than elsewhere. But the control exerted by structure is not absolute. Slope is more important at most sites. Gutters and grooves on Wudinna Hill and Ucontitchie Hill, northwestern Eyre Peninsula, South Australia, and in Plateau de Andohariana, Andringitra Massif, Madagascar (Vidal Romaní, Ramanohison and Rabenandrasana, 1997) follow along joints for short distances,

Figure 8.18. Kluftkarren (a) at Devil's Marbles, Northern Territory, Australia; (b) at Paarlberg, near Cape Town, South Africa; note flared basal wall in foreground, where regolith formerly lapped up against the bedrock.

but run in fresh, massive rock for the most part, even though joints are present nearby (Fig. 8.19). Some run diagonally across major joints. Also, on Wudinna Hill, some grooves have been eroded preferentially in the porphyritic granite that comprises the greater part of the residual, rather than in locally developed aplitic veins and lenses, though they have preferentially exploited the junction between the two rock types, and where this has occurred, the gutters are bordered by raised rims of aplite. Thus, although there is a general structural control of distribution of grooves in some areas, most are developed independently of structure in detail. Most workers have concluded that grooves and gutters are due partly to chemical and partly to mechanical processes. Debate on the origin of the channels has revolved around the relative significance of mechanical abrasion and chemical weathering, particularly by moisture, and around the role of biota, especially such plants as lichens.

Branner (1913) noted and investigated flutings, some of them 2 m deep, developed on granite and syenite residuals in Brazil. He attributed them largely to the mechanical action of running water. Others have also emphasised mechanical work. Klaer (1956), for example, considered both mechanical and chemical processes, but because both bed and sidewalls of the channels are smooth, emphasised abrasion by running water. Tschang (1961) suggested that the Pseudokarren of Pulau Ubin are due to wash.

Others, such as Ule (1925), attribute grooves wholly to chemical weathering. Scholz (1947, p. xlix) noted runnels on granitic inselbergs at Vredenburg and Witteklip, in the Western Cape Province of South Africa, and thought them due to "*solvent action of downward trickling solutions charged with cyclic salts and organic acids…*". The latter, he thought, derived from soil-filled rock basins developed in the upper slopes of the residuals. Support for such an origin involving the dominance of the chemical action of water comes from the development of runnels on the faces of blocks and boulders that generate insignificant volumes of runoff; from their occurrence on the interior walls of hollows and on overhanging faces draining very small areas, where the water adheres to the steep slopes by surface tension (Figs 8.17c and f) and from pitting found in the beds of several of the channels (Fig. 8.20).

198 *Landforms and Geology of Granite Terrains*

(a)

(b)

Figure 8.19. Gutters and flutings following and diverging from fractures, Wudinna Hill, Eyre Peninsula, South Australia: (a) general view, showing runnels draining a soil patch on upper slope (decantation runnels), (b) detail of part of Fig. 8.19a.

Most workers have steered a middle course and attributed runnels to both mechanical and chemical agencies. The weight of evidence favours chemical processes and especially solution. Flows of water cause the rock to be wetted in linear zones extending downslope; chemical reactions take place; subsequent flows of water remove the particles loosened by weathering and so a linear depression or channel develops. Once formed, the channel tends to gather water, and hence the wetting and erosional processes are augmented.

As for biotic effects, several agents and processes have been noted. Alexander (1959), for instance, considered lichens to be all-important. She argued that lichens colonise moist linear zones and that, by

Figure 8.20. (a) Pitted floor of runnel on eastern slope of Ucontitchie Hill, northwestern Eyre Peninsula, South Australia. (b) Potholes along grooves on Pic Boby, Andringitra Massif, Madagascar.

retaining moisture, they contribute materially to channel development. Lichens are capable of both mechanical and chemical weathering, but it is surely more likely that they colonise zones that are already moist rather than growing in linear patterns in the first place. For this reason, most workers assign to lichens, mosses and other organic agencies a contributory rather than dominating or initiating role. Moreover, in some circumstances organic (algal) slime has a protective function. Rills rich in organic matter initially increase the rate of weathering and give rise to intensely pitted channel floors.

Some channels are merely continuations of the gutters draining upper slopes, but some are confined to the steep slopes. Pot-holes indicate that scouring and abrasion are active (Fig. 8.20b), but the very steepness, and, in places, the overhanging character of some of the bounding walls of the granite residuals preclude simple abrasion as a significant factor because the flows become separated from the bedrock surface (Vidal Romaní, Ramanohison and Rabenandrasana, 1997; Petit, 1971).

In these circumstances a number of other factors may be cited. Separation of flow may result in or lead to collapse and impact in linear zones – the so-called water-curtain effect. Scouring could cause increased surface roughness and induce further perturbation, air entrainment and turbulence, and hence increased erosion in linear zones running down steep bedrock surfaces. The field evidence suggests that, important as free flows are in evolution of these grooves, trickles of water are also significant. Such trickles, some deriving from patches of soil, others from sheeting joints, persist long after rain has ceased, even on overhanging slopes to which they adhere by surface tension. These are seepages rather than flows. The mechanical action of running water cannot be invoked in explanation of the grooves. Besides wetting the surfaces over which they flow and encouraging algal and other growths, such seepages and trickles transport salts and minerals in solution and are essential to the formation of speleothems (see Chapter 10).

Some have suggested that where high velocity flow develops, cavitation may occur, with high pressure waves and high speed water jets causing local rock shattering; and this could conveniently account for the pot-holes and other depressions found along some grooves, as, for example, on

Yarwondutta Rock, northwestern Eyre Peninsula, South Australia. Unfortunately, it is unlikely that the necessary high speed flows develop on the outcrops under review.

8.6.3 *Surface or subsurface initiation?*

The arguments concerning the relative importance of mechanical and chemical processes in the development of gutters are interesting *per se*, but are also linked with the question of their initiation, for there are strong indications that some of the forms, at least, originated at the weathering front.

In 1849 Logan reported deep grooves developed in granite on Pulau Ubin, an island at the eastern end of the Johore Strait (Fig. 8.21). He reported similar features from uplands in the southern part of the Malay Peninsula. On Pulau Ubin many of the granite blocks exposed on the beach were

Figure 8.21. (a) General view of fluted block at Pulau Ubin, Singapore. (b) Detail of fluting on nearby Frog Island (Pulau S'kodo or Sekudu).

fluted and grooved, and inland many vertical and near-vertical faces of the blocks were scored and grooved, some of them flask-shaped in cross-section, and some deep and narrow, being about 0.76 m deep and only 0.61 m across. For various reasons, Logan (1894) considered contemporary processes of weathering and erosion inadequate to explain the grooves, but he made one crucial observation and several comments germane to the present discussion. He noted that some of the grooves of Pulau Ubin extend beneath the soil cover. Unless there had been exposure, development of flutings, burial and subsequent exhumation, a sequence of events unsupported by any evidence Logan (1849) could observe, this would appear to demonstrate that the flutings had been initiated below the soil cover. Nevertheless, Logan (1849) remained uncommitted in his 1849 paper. By later (Logan, 1851), however, he asserted categorically not only that the grooves but all the boulders and blocks he had observed on Pulau Ubin together constituted what would today be called an etch surface or complex weathering front. Thus, Logan wrote:

"If ... we conceive the external layer of the island, when it first became exposed to decomposition, to have resembled in character the zone that has been laid open for our inspection ... it is easy to comprehend how the wasting away of the more decomposable parts might at last leave exposed masses, including bands of the less stubborn material already partially softened or disintegrated under ground, and that the action of the atmosphere and rain-torrents would gradually excavate the more yielding portions until the solid remnants exhibited their present shapes." (Logan, 1851, p. 328).

Logan (1849, 1851) was clearly advocating a subsurface origin, development at the lower limit of weathering or weathering front. Evidence from Eyre Peninsula and many other areas clearly demonstrates that many gutters have been initiated there. Some extend downslope on quite steep inclines and become flutings. Many grooves and gutters in granite are, as Logan (1849, p. 6) put it, *"prolonged beneath the ground"*. There is no suggestion of burial and exhumation following development. On the contrary, it is grus, or weathered granite, which has been excavated to reveal the scored weathering front. These forms are therefore initiated beneath the soil surface, and have been merely enlarged and modified after exposure.

On recently, artificially-exposed, weathering fronts the gutters became wider and shallower, less well-defined, as they extend below the soil level (Fig. 8.22a). This is because the water

Figure 8.22. Flutings shallower and less well-defined (a) at Tampin, West Malaysia.

Figure 8.22. (b) In the Traba Massif, Galicia. (c) Flutings with a flask-shaped cross-section, Cassia City of Rocks, Idaho, USA.

percolating along the front has to pass between rock particles so that the flow becomes diffuse. Flutings that tend to lose definition downslope, and are therefore probably of subsurface origin, can be seen on blocks, boulders and hillslopes in many parts of the world (e.g. Fig. 8.22b). Some are flask-shaped in cross-section due to the induration of the rock surface (Fig. 8.22c).

In general, however, the opposite view has prevailed, finding expression for example in Wilhelmy's (1958, p. 199) assertion that flutings in the humid tropics evolved "*auf freiliegenden Gesteinsoberflächen*". And some flutings in granite, like those scored on the menhir of St Uzek, in Brittany, France (Lageat, Sellier and Twidale, 1994) are manifestly subaerial forms, having developed during the last 5,000–5,500 years, since the erection of the monolith (Fig. 8.23). Watson and Pye (1985) advocate an epigene origin for Karren in Swaziland, but while stating that there is no evidence for a subsurface origin of these (and other) minor forms, they do not discuss whether

Minor forms developed on steep slopes 203

Figure 8.23. Menhir at St Uzek, northern Brittany, France, showing flutings developed since erection. The basin (X) shows that the vertical face on which it is located was originally horizontal.

the residuals that are host to the features are two-stage forms or not. If they are, then the surfaces on which the Karren occur were once at the weathering front.

8.6.4 *Inversion*

At several sites, but especially well-developed at Wave Rock, Western Australia, and at Turtle, Pildappa and Yarwondutta rocks, on northwestern Eyre Peninsula, Southern Australia are former gutters that are now ribs: there has been local inversion of relief (Fig. 8.10a and Fig. 8.24). Most gutters draining from the flattish upper surfaces of hills continue down the steeper marginal slopes as channels scored in the lower marginal slopes (Twidale, 1962). On the very steep slopes – even those that are overhanging – surface tension allows seepages to adhere to the channel floor, though higher velocity flows separate from the bedrock, in places causing minor plunge pools to form in the channel floor (Figs 8.10a and b). Some gutters drain debris-filled vegetated basins. They carry blue-green algae (on Eyre Peninsula, *Calothrix spp.*), which coat slopes and channel floors.

Though thin, these coatings are, on one interpretation, evidently enough to protect the bedrock, and on overhanging slopes facing north or northwest later flows have differentially eroded the unprotected zones adjacent to the original flutings, leaving the old channel floors in relief, as ribs. There are complications, for the later lateral channels are not coated with algal remains.

204 *Landforms and Geology of Granite Terrains*

Figure 8.24. Inverted Rillen on southern, flared sidewall of Turtle Rock, near Wudinna, Eyre Peninsula, South Australia.

Environmental changes could be invoked to account for the absence of algae in the later flows, but not all ribs are associated with rock basins. Indeed some on Yarwondutta Rock serve very small catchments. It may be that, on leaving the confined gutters of the upper slopes, the small streams spread on to the upper parts of the flared overhangs. Algae continue to grow on the central moister areas, but the wash lateral to these causes erosion of the flaked or laminated surface rock, creating depressions and leaving the central area in relief (Twidale and Campbell, 1986).

But other explanations are possible. For example, the colonisation of local rises by algae can be seen as effects rather than causes. It can be suggested that the ribs are due to the contrast in rate of weathering between wet areas and those that are alternatively wet and dry. The surface of the residuals is flaked or scaled, inherited either from the weathering front (Chapter 3), or developed after exposure by insolation or by water-related weathering. The linear vertical zones downslope from the gutters scored on the upper surface of the inselbergs are moist because of the water flowing from above, and retained in the interstices between flakes. Summer storms maintain this supply in these narrow sectors. To either side there are zones which are drier in summer (in southern Australia) because the supply of moisture drawn up between flakes by capillarity from the soil at the base of the hill disappears as the surface soil is desiccated. Summer storms replenish the supply here also, but the soils are well below field capacity and the meteoric waters percolate to the water table at depth. Alternations of wetting and drying cause the flakes to fall away, so that those sectors of the sidewall are worn back, leaving the more stable zones below the gutters in relief as upstanding narrow ribs. Such wetting and drying of basal slopes could also explain some of the basal fretting referred to earlier (Chapter 8.2).

REFERENCES

Alexander, F.E.S. 1959. Observations on tropical weathering: a study of the movement of iron, aluminium and silica in weathering rocks at Singapore. *Quarterly Journal of the Geological Society of London* 115: 123–142.

Anderson, A.L. 1931. Geology and mineral resources of eastern Cassia County, Idaho. Idaho Bureau of Mines and Geology Bulletin 14.

Branner, J.C. 1913. The fluting and pitting of granites in the tropics. *Proceedings of the American Philosophical Society* 52: 163–174.

Bulow, K. 1942. Karrenbildung in kristallen Gesteinen? *Zeitschrift Deutsche Geologische Gesellschaft* 94: 44–46.
Carlé, W. 1941. Karrenbildung im Granit der galicischen Kuste bei Vigo (Norwestspanien). *Geologie Meere und Binnegewasser* 5: 55–63.
Clarke, E. de C. 1936. Water supply in the Kalgoorlie and Wheat Belt regions of Western Australia. Proceedings of the Royal Society of Western Australia 22, xi–xliii.
Clayton, R.W. 1956. Linear depressions (Bergfussniederungen) in savannah landscapes. *Geographical Studies* 3: 102–126.
Cotton, C.A. 1942. Climatic accidents in landscape making. Whitcomb & Tombs, Wellington.
Dumanowski, B. 1960. Comment on origin of depressions surrounding granite massifs in the eastern desert of Egypt. *Bulletin of the Polish Academy of Sciences* 8: 305–312.
Hellstrom, B. 1941. Några iakttagelser över vitting, erosion och slambildning i Malaya ich Australien. *Geografiska Annaler* 23: 102–124.
Jutson, J.T. 1914. An outline of the physiographical geology (physiography) of Western Australia. Geological Survey of Western Australia Bulletin 61.
Klaer, W. 1956. Verwitterungsformen in granit auf Korsika. Petermanns Geographischen Mitteilungen Ergännzungsheft.
Lageat, Y., Sellier, D. and Twidale, C.R. 1994. Mégalithes et méteorisation des granites en Bretagne littorale (France du nord-ouest) *Géographie Quaternaire et Physique* 48: 107–113.
Logan, J.R. 1849. The rocks of "Palo Ubin". Genootschap van Kunsten Wetenschappen (Batavia) 22: 3–43.
Logan, J.R. 1851. Notices of the geology of the Straits of Singapore. *Quarterly Journal of the Geological Society of London* 7: 310–344.
Mueller, J.E. and Twidale, C.R. 1988. Geomorphic development of City of Rocks, Grant County, New Mexico. *New Mexico Geology* 10: 74–79.
Petit, M. 1971. Contribution à l'etude morphologique des reliefs granitiques à Madagascar. Société Nouvelle de l'Imprimerie Centrale, Tananarive.
Petit, M. 1990. Géographie Physique Tropicale. Approche aux Etudes du Milieu, Karthala ACCT.
Ruxton, B.P. 1958. Weathering and subsurface erosion in granite at the piedmont angle, Balos, Sudan. *Geological Magazine* 45: 37–44.
Schmidt-Thomé, P. 1943. Karrenbildung in kristallinen Gesteinen. *Zeitschrift Deutsche Geologische Gesellschaft* 95: 53–56.
Scholz, D.L. 1947. On the Younger Pre-Cambrian granite plutons of the Cape Province. *Transactions of the Geological Society of South Africa* 49: xxv-lxxxll.
Tschang, H.-L. 1961. The pseudokarren and exfoliation forms of granite on Pulau Ubin, Singapore. *Zeitschrift für Geomorphologie* 5: 302–312.
Twidale, C.R. and Campbell, E.M. 1986. Localised inversion on steep hillslopes: gully gravure in weak and resistant rocks. *Zeitschrift für Geomorphologie* 30: 35–46.
Twidale, C.R. 1962. Steepened margins of inselbergs from north-western Eyre Peninsula, South Australia. *Zeitschrift für Geomorphologie* 6: 51–69.
Twidale, C.R. 1967. Origin of the piedmont angle, as evidenced in South Australia. *Journal of Geology* 75: 393–411.
Ule, W. 1925. Quer durch Sudamerika. Quitzow, Lubeck.
Vidal Romaní, J.R., Ramanohison, H. and Rabenandrasana, S. 1997. Géomorphologie granitique du Massif de l'Andringitra: sa relation avec l'évolution de l' Ile pendant le Cénozoïque. *Cadernos do Laboratorio Xeolóxico de Laxe* 22: 183–208.
Watson, A. and Pye, K. 1985. Pseudokarstic micro-relief and other weathering features on the Mswati Granite (Swaziland). *Zeitschrift für Geomorphologie* 29: 285–300.
Wilhelmy H. 1958. Klimamorphologie der Massengesteine. Westermanns, Brunswick.
Wilhelmy H. 1964. Cavernous rock surfaces (tafoni) in semi-arid and arid climates. *Pakistan Geographical Review* 19: 9–13.

9

Minor forms developed on gentle slopes

Some bornhardts and large residual boulders have gently sloping upper surfaces. Some of the crestal bevels of bornhardts were platforms before the plains around them were lowered. They are commonly dimpled, due to the development of rock basins, and grooved by clefts and runnels. In addition, though more rarely, pedestals, rock doughnuts and rock levees are developed.

9.1 ROCK BASINS

9.1.1 *Description*

Rock basins are depressions formed in solid bedrock (Twidale and Corbin, 1963). Morphologically they vary in detail, but most are oval, elliptical or circular in plan. Some, strongly influenced by jointing, are angular in form, and others, resulting from the coalescence of two or more individuals, are irregularly lobate in plan. All are temporary water storages. Water accumulated from rainfall and runoff is lost either through evaporation, or use by animals (including humans), or by underground seepage. Alternatively, if supply exceeds storage and losses, the water in the basin overflows, cutting an outlet so that the depression eventually becomes part of an integrated drainage system.

Several morphological types have been recognised:

- Pits are hemispherical, are developed on gentle slopes and are isolated. Outlets are rare (Fig. 9.1a). Pits with overhanging sidewalls are called flasks.
- Pans are also developed on gentle slopes, and are comparatively shallow and flat floored (Figs 1.2f, 9.1b and c). Outlets are few, though some pans are linked by gutters or runnels, and thus form part of an integrated drainage system (Fig. 9.2a).
- Armchair-shaped hollows are asymmetrical in section normal to the contours. They have high backwalls on the upslope side, but are open downslope (Figs 9.1d and e), where there is frequently a distinct outlet in the form of a runnel or gutter. Armchair-shaped hollows are typical of the moderately steep (20–30°) slopes that lead down from the flattish crestal areas to the steep bounding slopes and the plains (Fig. 9.2b).
- Cylindrical hollows vary in plan shape, though they are approximately circular or oval, and are rectangular in vertical section, so that they are appropriately referred to as being of cylindrical or terete form (Fig. 9.1f). The bedrock cylinder opens into a sheet fracture so that there is in some instances a through drainage, and the shallow spiral scallops discernible on some sidewalls reflect the swirl of waters rushing down the opening during and after heavy rains; more commonly the basins are choked by detritus and vegetation, so that the bedrock morphology is only revealed by excavation and the removal of the fill.

Rock basins vary in size as well as shape. The modal diameter is of the order of a metre and of depth perhaps 0.5 m. Most hold a few litres of water, but some are much larger. One on King

208 *Landforms and Geology of Granite Terrains*

Figure 9.1. (a) Pit on Pildappa Rock, northwestern Eyre Peninsula, South Australia. (b) Pan on Kulgera Hills, Northern Territory. (c) Broad shallow pans on Paarlberg, near Cape Town, South Africa.

Minor forms developed on gentle slopes 209

Figure 9.1. (d) Armchair-shaped hollows on flanks of Pildappa Rock, northwestern Eyre Peninsula, South Australia. (e) Large single armchair-shaped hollow on The Humps, southern Yilgarn, Western Australia. (f) Cylindrical hollow on the Kwaterski Rocks, northwestern Eyre Peninsula, South Australia.

210   *Landforms and Geology of Granite Terrains*

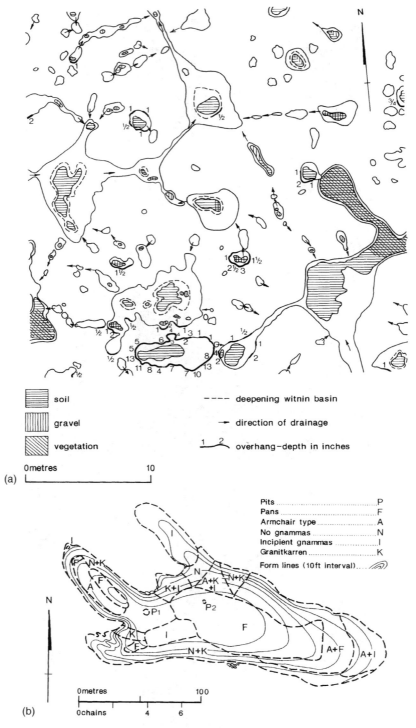

Figure 9.2. (a) Plan of part of upper surface of Pildappa Rock, showing basins and gutters. (b) Plan of Pildappa Rock, northwestern Eyre Peninsula, South Australia, showing distribution of types of rock basin.

Rocks, near Hyden, in Western Australia, has a capacity of almost 7 million litres (1.5 million gallons). The various types have been described from many parts of the world. Basins are widely distributed and are characteristic of granite outcrops of all shapes and sizes.

9.1.2 Nomenclature

Rock basins have been referred to by several names. Though first mentioned in the scientific literature in the late Eighteenth Century, the forms had actually been perceived, measured and discussed in general terms for many centuries. They were described from Dartmoor in southwestern England, where they were known as rock basons, in 1291, more than seven centuries ago (Worth, 1953).

Rock basins of various types have been referred to also as rock holes, weather pits, pot-holes, water eyes, cauldrons, cisterns, granite pits, and, in Australia, night-wells, and gnammas. In Idaho they are known as bath tubs, and other local names have doubtless also been applied. They are the tanques and vasques rocheuses of the French literature, and the Verwitterungsnäpfe, Opferkessel, Baumverfallspingen, Felsschüssel and Dellen of German workers. They have also been called kociolki in Poland, pias, marmitas, cassolas and pilancones in Spain, caldeiros, poços, and oriçangas in Brazil, and araceenhorst in Surinam.

9.1.3 Origin

For some years, about two centuries ago, rock basins were attributed to human agencies, and in Britain, for example, were thought to be related to Druidical ceremony. Red staining of the water held in the basins and due to red algal growths, or of the basin rims (due to the release of iron during the alteration of biotite and other minerals) suggested blood to the ancients. Such an anthropogenic origin can be dismissed as a general explanation if only because many basins were initiated beneath the natural land surface (Twidale and Bourne, 1975), at the weathering front (see below).

Water, with solution, hydration and hydrolysis the major processes at work, is responsible for differential weathering at the weathering front and also for growth of the basins after exposure (Smith, 1941 and see Fig. 9.3a). The granite is rotted and disintegrates, leaving a residue of gritty sand, with some larger fragments (Fig. 9.3b). Evorsion, and mechanical abrasion by granite sand carried by swirling water, however, contribute to the formation both of armchair-shaped hollows located on the flanks of residuals, and of the cylindrical forms.

It is difficult to ascertain with certainty why specific basins are located where they are, for the rock that provides the evidence has been weathered and evacuated. Any condition that can produce a depression, however minor, is enough, for accumulated water alters the rock with which it comes into contact and so enlarges the original hollow. Granite is not compositionally homogeneous. Xenoliths are commonplace and frequently induce preferential weathering as do clusters of phenocrysts (Fig. 9.3c). Thus, in the Linares district of Andalusia, and at several sites in Galicia and Girona, northwestern and northeastern Spain respectively, small masses of biotite-rich or otherwise melanocratic rock have been exploited to produce small, angular basins comparable in size and shape to the xenoliths that are still intact. Similarly, the essential minerals in granite are not everywhere evenly distributed. It is not uncommon to find bands or discrete masses of feldspar or biotite, both of which minerals, being susceptible to moisture attack, could be preferentially weathered to form an initial hollow that could later develop into a basin.

Gravitational loading (Fig. 9.4), i.e., the pressure exerted by the rock itself when the load concentration effect is produced (minimum thickness calculated to be of the order of 300 m), could cause local strain, lattice distortion and disequilibrium, and fragmentation of crystals. In contact with moisture, either at the surface, or, more probably, initially within the zone of groundwater, the strained sites or linear zones would be weathered, inducing the development of small depressions. Similarly, tectonic strains may be concentrated in pressure points (like those induced by the insertion of rock bolts) as a result of uneven contacts between blocks beneath the surface, as for example

212 *Landforms and Geology of Granite Terrains*

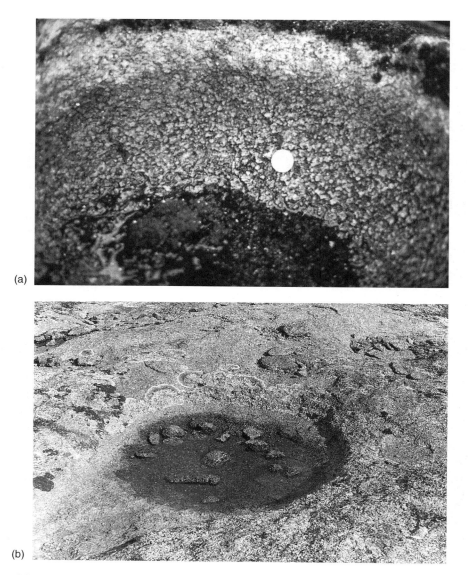

Figure 9.3. (a) Subaerial differential weathering of minerals of the side wall of a pit developed on granodiorite from Monte Louro (Galicia, northwestern Spain), showing pitted surface (see pp. 71–73). (b) Shallow saucer-shaped pan on Yarwondutta Rock, northwestern Eyre Peninsula, South Australia, showing large fragments remaining in pool after being released by the weathering of the surrounding granite to form grus or granite sand.

along wavy planes of dislocation, and with results similar to those induced by rock plastification and consequent late weathering (Vidal Romaní, 1989).

Many basins have formed along or at the intersection of fractures and fissures, and some are elongated along the fracture plane (Figs 9.5a and b; also Fig. 9.2a). Others, developed in structurally complex country rock, have irregular outlines as a result of the exploitation of various fractures (Fig. 9.5c). The original depression may be irregular in shape, but moisture attack tends to produce rounded forms of considerable perfection because of the concentration of weathering, whether by water or by frost action, on projections of rock.

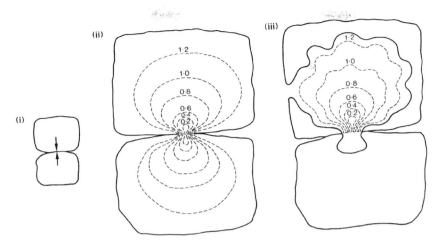

Figure 9.4. (i) Schematic evolution of tafoni and basin in two blocks in contact along a fracture, (ii) load concentrations, and (iii) the resultant forms. The figures indicate stability coefficient (see Vidal Romaní, 1989).

Figure 9.5. (a) Elongate pit formed along fracture, Ebaka Dome, French Cameroon (M. Boyé).

Several workers have argued that water collecting in chance crevices and depressions in exposed rock surfaces could enlarge those initial depressions to form rock basins. Hedges (1969) went so far as to state that "They are never found to be developed beneath a soil cover in road cuts and other excavations". Watson and Pye (1985) make similar assertions regarding the basins developed on the Mdzimba Hills in Swaziland, though, as with flutings or grooves (see Chapter 8.6), it is relevant to consider the origin of the host landform.

Some rock basins, like that developed on the upper surface of the artificially-erected menhir at St Uzek in Brittany (Lageat, Sellier and Twidale, 1993 and see Fig. 8.23), are undoubtedly of epigene origin. But many rock basins (like other granite forms, major and minor) are demonstrably initiated in the subsurface. Some basins have been revealed by excavations of road cuts and of dam foundations in northern Spain. Similar evidence has been revealed at sites at Midrand, between

Figure 9.5. (b) Elongate basin or bathtub, Childara Rocks, eastern Victoria Desert, South Australia. (c) Irregular pan, Buccleugh, near Johannesburg, South Africa.

Johannesburg and Pretoria, in South Africa and at the Ebaka Quarry in southern Cameroon, west Africa (Twidale, 1988).

Northwestern Eyre Peninsula, South Australia, receives only a moderate and unreliable rainfall, and considerable attention has been devoted to water conservation. Over the past eighty years, many small reservoirs have been excavated at the lower margins of the granite residuals. Elsewhere, the weathered granite or grus has been stripped away, exposing the weathering front in

Minor forms developed on gentle slopes 215

Figure 9.6. (a) One of the Kwaterski Rocks, northwestern Eyre Peninsula, South Australia, with saucer-shaped depressions on recently cleared platform. (b) Another of the Kwaterski Rocks, showing depressions and shallow fracture-controlled cleft or Kluftkarren.

platforms that serve as catchments. On exposure, the weathering front is seen to be dimpled and grooved due to the development, beneath the regolith cover, of shallow, saucer-shaped depressions and linear channels (Figs 9.6a and b). At some sites many small, comparatively shallow, pits have developed in close proximity to one another, so that the basins are separated by knife-edged ridges. Together they look like a choppy sea, or the surface of a meringue.

Figure 9.7. Pan with large overhanging sidewall, Yarwondutta Rock, near Minnipa, northwestern Eyre Peninsula, South Australia.

9.1.4 Differentiation of major types

Whether initiated on exposed surfaces or at the weathering front, the original saucer-shaped depressions are further developed and several morphological types evolve on exposed surfaces. Because water is in contact longer with the floors and lower sidewalls of the depressions than with the upper sidewalls, weathering effectively causes the basin to be extended vertically into the rock mass and, depending on whether or not there was undermining of the surface layer, a pit or a flask-shaped depression is formed.

Where water acts on sensibly homogeneous granite, the availability of water (runoff), capacity of the depression to hold water and the duration of wetting are all-important, and hemispherical pits form as a result of reactions between water and the bedrock. The most common type of basin, found in various climatic and lithological settings, is the flat-floored pan. Pans are circular or elliptical in plan and their outlines are generally smooth and regular. Most are 1–2 m in diameter and some 30–50 cm deep, though composite features are, of course, larger and also less regular in plan. The sidewalls of many pans are overhanging by a matter of a few centimetres, though on Yarwondutta Rock, northwestern Eyre Peninsula, South Australia, the floor of one basin extends about one metre beneath the edge of the protruding lip (Fig. 9.7). Pans are consistently developed in flaggy rock characterised by discontinuous fractures that run essentially parallel to the rock surface. Water readily penetrates along these joints so that lateral extension takes place more rapidly than does vertical growth. This explains the large diameter/depth ratio of the pans. The floors of the pans are frequently coincident with parting planes. This argument finds support at many sites. At Lightburn Rocks, in the eastern Great Victoria Desert in South Australia, for example, pans are well-developed on the crestal areas of flaggy or laminated rock, but pits occur on the gentle, lower slopes of the flanks where massive rock is exposed.

The overhanging sidewalls of the pans are due to three effects, one structural, one biotic and the other related to duration of wetting. First, in many areas, a superficial induration of oxides of iron and manganese is developed, and this effectively cements the bedrock so that it withstands weathering and erosional attack better even than does the fresh rock. Second, lichens do not colonise surfaces which are frequently under water, but they are widely established on the upper slopes of rock basins. Third, water persists longer in the floor of the depression and the lower sidewalls are, for this reason, weathered back more rapidly than are the higher zones. This last factor is everywhere

Minor forms developed on gentle slopes 217

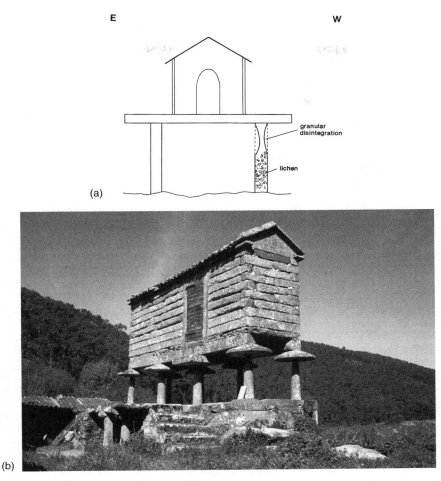

Figure 9.8 (a) Diagram showing hórreo or maize-storage house on stilts, and the weathering of part of granite stilt not colonised by lichens. (b) Galician hórreo from Laxe, A Coruña, northwestern Spain.

important, for overhanging sidewalls are developed in areas, such as Dartmoor, southwestern England, and the Yosemite region of the Sierra Nevada, California, where there is no detectable surface induration (but where there are lichen colonies).

The role of lichens is difficult to assess (Scott, 1967; Fry, 1926). It has been shown that lichens are agents of weathering. Their hyphae penetrate the rock and assist in physical disintegration. They also facilitate water entry and in addition extract selected minerals for their own use. On the other hand, and without denying that lichens cause weathering, there is observational evidence that they protect rock surfaces. They not only stabilise the surface in an absolute sense but the rate of weathering is greater on adjacent exposed or non-colonised surfaces. For example, in many parts of northern Spain small houses on stilts called hórreos are constructed for the storage of corn (maize) heads. The house sits on a platform which is much wider than the spacing of the stilts, the overhang preventing rodents climbing up to the house and the maize (Figs 9.8a and b). Such an hórreo was built of the local stone, granite, at Louro, A Magdalena, near the Ría de Muros, in southern Galicia in 1950. Within 50 years lichens had colonised those (lower) parts of the stilts exposed to rain from the west, but the upper parts, in the shade of the platform, were not colonised. The exposed granite clearly suffered disintegration, whereas the areas covered by lichens remained

218   *Landforms and Geology of Granite Terrains*

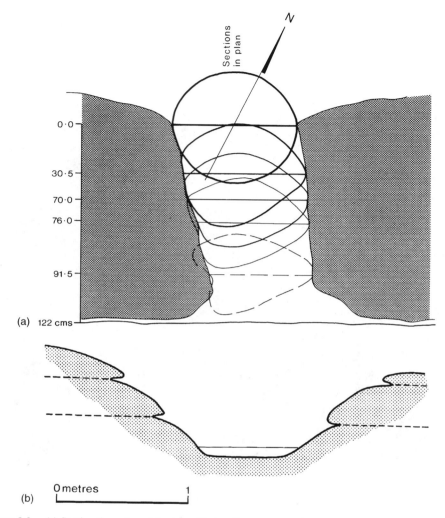

Figure 9.9. (a) Section through rock basin on Kestor, Dartmoor, southwestern England. (b) Section through cylindrical hollow, Kwaterski Rocks, northwestern Eyre Peninsula, South Australia.

intact. The site is near the sea, so that salt weathering may be involved, but the protective role of lichen is clearly demonstrated.

The topographic distribution of various types of basin is consistent with the suggested explanations of the major morphological types (Fig. 9.2b). Pans are most numerous on broadly rolling upper surfaces or platforms, though armchair-shaped hollows are prominent on the steeper marginal zones. Pits occur in topographic lows, below the level of flaggy or laminated rock on which pans are so profusely represented, and in homogeneous bedrock.

This explanation may be of general application, for only pans are developed in southwestern England where the superficial zone of the granite outcrops, on which the rock basins were developed, is flaggy. The basin on Kestor for example, has penetrated through two thick laminae so that its cross section profile is irregular (Fig. 9.9a). The general point was appreciated more than a century ago by Jones (1859), who wrote that the tabular (or flaggy) structure of the Dartmoor granite was "*probably the cause of frequent occurrence of basins with flat bottoms*".

Where basins develop in granite that is sensibly isotropic, but where there is superficial mineral induration, or biotic protection, flask-shaped basins or Opferkessel develop. Where the pits have penetrated through the slab or sheet in which they have developed to the sheeting joint beneath, water draining into it runs along the basal fracture so that the previously enclosed basin becomes a throughway for running water. The base is widened, the water swirls around, creating shallow grooves or scallops in the rock walls and a cylindrical hollow is formed (Figs 9.1f and 9.9b).

9.1.5 Evacuation of debris

The evacuation of debris from basins has given rise to some discussion. It is to some extent a non-question, for much of the granite sand remains in the basins, where, because it retains moisture and attracts vegetation and hence organic acids, it probably enhances the rate of basin development. Also, organic growths in the basins cause overflows of water rich in humic acids and also are favourable to the growth of blue-green algae, which may affect the development of flutings or Rillen on steep slopes (see Chapter 8).

Many rock basins, both in Australia and elsewhere, have been cleared of debris by humans. Not only did Aboriginal people clean out the basins in order to make them more effective as water storages, but Afghan cameleers transporting stores to inland stations, and shepherds and stockmen, have done the same. Slabs of rock (and later, sheets of corrugated iron) were placed over basins in an attempt to reduce evaporation and contamination. Such anthropogenic cleansing may be not so widespread as has been supposed, because physical and chemical agencies, i.e., solution, suspension, etc. have also been invoked to account for clean basins in humid areas where rock basins are very common features.

Some workers give credence to the deflational work of wind in suitable environments, and even in parts of humid western Europe, coarse fragments are concentrated by the evacuation of fines. Dissolved salts are evacuated in solution. Also, fines are carried in suspension by turbulent flow, and water flushing through the drainage network seems to offer an adequate and feasible mechanism to account for the transport of clays and fine sand derived from the weathering of the granite. The layering of sediments according to grain size within some basins suggests that turbulence is not significant everywhere.

Sediment left behind in the basins shows concentrations of coarse debris in the centre of the depressions and even small ripple marks; this in contrast with lower energy regimes where coarse debris, fallen from the sides of the depression, is found at the edges of the basin deposit.

9.1.6 Rate of development

The evidence concerning the rate of growth is equivocal. Rock basins have developed on the flattish upper surfaces of dislodged blocks in some areas suggesting that the forms evolve in a matter of a few thousand of years. Basins are widely and well-developed in several polar regions (e.g. Scandinavia, the Canadian Arctic) that were covered by ice until a few thousands of years ago. The implication is that the basins have either survived glaciation or have developed since the exposure of the surfaces on which they occur. The latter is not impossible. In the Bohemian Massif basins have evidently developed at a rate of several centimetres each century, and, though the measurements were not as precisely controlled as might be thought desirable, there is strong suggestion of rapid growth. On the other hand, in the Iberian Peninsula and in the Snowy Mountains of New South Wales, basins are absent from glaciated areas but are found on adjacent, unglaciated, surfaces. The state of the local glacier ice (hot, cold) is crucial, and the existence of preglacial, as well as postglacial, basins cannot be ruled out.

On the other hand, Mistor Pan, which is a large and well-known rock basin on Dartmoor, southwestern England (Worth, 1953), was described in 1291 and again in 1609. Its depth was measured in 1828, 1858, 1875 and 1929. All results were similar, the differences being as readily accountable in terms of the difficulty in determining the upper limit of the depression as by any real increase in depth. Thus, whereas in some regions, basin development has been rapid, elsewhere it is negligible.

220   *Landforms and Geology of Granite Terrains*

Climatic factors may play some part but the structure and condition (fresh, weathered) of the granite also influence the rate, as well as the type, of development.

9.2.   PLINTHS AND ASSOCIATED BLOCKS AND BOULDERS

9.2.1   *Description*

Plinths are low, smooth-topped projections that stand a few centimetres or tens of centimetres above the level of the adjacent rock surface. They are commonly surmounted by a boulder or block of rock (Figs 5.11 and 9.10). Some occur on slopes of gentle or moderate inclination, but others

Figure 9.10.   Granite plinth, with boulder (a) on Domboshawa near Harare, Zimbabwe. (b) Low on eastern flank of Ucontitchie Hill, Eyre Peninsula, South Australia.

stand above the general level of sensibly flat platforms. Examples have been noted at Domboshawa, in Zimbabwe, in several parts of South Australia and Western Australia, and in the Serra do Gêrez, on the border between Portugal and Galicia, NW Spain.

9.2.2 Origin

Similar, though more dramatic, pinnacles due to the deep dissection of weak, unconsolidated or intensely weathered materials have been described from various parts of the world. They are commonly attributed to the protection of the preserved acicular columns by the slabs and blocks that cap them. Similarly, in the Serra do Gêrez (Coudé-Gaussen, 1981), glacial erratics afford protection to the underlying granite surfaces, and also demonstrate that the plinths in question must have developed since the last glacial retreat in the area, that is, in the last 14,000 years.

Protection by the block or boulder resting on the plinth also seems to explain the latter forms. Assuming first that the block or boulder rests on a bare rock platform, water dripping from the edges of the protective residuals forms pools and depressions that weather the adjacent platform, as well as spreading laterally and cutting into the basal slope of the plinth. In this way, the adjacent surfaces have been lowered and once the depressions are linked, washing from the slope courses through them, lowering the surface still further, and leaving the protected plinth in relief.

The effect of drip pools is convincingly demonstrated at Tolmer Rocks, in the Upper South East district of South Australia (Fig. 9.11). There, water dripping from a large residual boulder has, through excessive wetting and weathering of the platform on which it falls, caused the development of a series of shallow pools, the pattern of which reflects exactly the plan shape of the boulder. The pools have extended laterally and merged to form a narrow, shallow moat around the base of the boulder. Any runoff from upslope washes into the moat. Thus, the moat is deepened and runoff from upslope is diverted. In this way, that area of the platform located beneath the boulder is not attacked by running water as it is the immediately adjacent exposed surface. Weathering proceeds beneath the perched boulder, and the surface of the pedestal is also affected (see Chapter 10), but this is slow compared with the effects of weathering and wash on the slope as a whole.

The same mechanism is adequate to explain plinth development on platforms with a regolithic cover, for the water dripping on to the surface would infiltrate the cover and cause the weathering front to penetrate deeper forming an annular depression mimicking the plan shape of the boulder rim.

Figure 9.10.   (c) On Amboromena Pic, Andringitra Massif, Madagascar.

222 *Landforms and Geology of Granite Terrains*

Figure 9.11. (a) Tolmer Rocks, South East district, South Australia, showing part of large residual boulder, moat and plinth. (b) Section showing suggested mode of development of moat and plinth.

9.3 PEDESTAL ROCKS

9.3.1 *Terminology*

Pedestal rocks have been reported developed in granite, sandstone, dolomite, rhyolite and basalt (Twidale and Campbell, 1992). They are widely distributed and occur in most if not all climatic contexts. Pedestal rocks are mushroom-shaped pillars, consisting of a cap or table supported by a stem or shaft (Figs 8.13 and 9.12). They are also known as mushroom rocks, hoodoo rocks, Pilzfelsen, Tischfelsen, roches champignons and rocas fungiformes. They differ from the plinth and boulder assemblage described earlier in that pedestal rocks are essentially a coherent whole, whereas a plinth is separated from the block or boulder by an open, frequently gaping, fracture. The boulder may be unstable (a balancing rock, balanced rock or logging stone – Fig. 5.11d), a condition not found in pedestals.

9.3.2 *Origin*

The cap may be structurally resistant but this is not a necessary condition. Pedestal rocks have been variously attributed to sandblasting by ice crystals which may be correct in hyper-arid warm and cold deserts, though the wind almost certainly exploits bedrock weaknesses due to weathering

Minor forms developed on gentle slopes 223

Figure 9.12. (a) Mushroom rock in dolomite, with massive cap and flared sidewall, and (b) cleft in dolomite pavement showing flared sidewall, at Ciudad Encantada, near Cuenca, central Spain.

(see Fig. 8.13b), and to river erosion. The presence of an abundance of moisture and resultant weathering near (but above) ground level has also been invoked in explanation of pedestal rocks. In that case, pedestal rocks ought to be consistently asymmetrical because of insolational drying, but they are not. Also, if moisture were responsible, more is retained in the regolith than on bare rock surfaces. Shallow subsurface alteration can be expected to exceed that on outcrops. In addition, biota live in the soils and organic acids are formed there. The most probable explanation for pedestal rocks is that the cap, structurally more resistant or not, is exposed, and is on that account weathered less rapidly than the shaft which is in continuous contact with moisture retained in the regolith. This suggestion finds support in the development of pedestals with flared sidewalls in dolomite at the Ciudad Encantada, near Cuenca in central Spain, and also by the development of such features in clefts weathered below the plateau, the level of the pedestal caprock, in the same area (Fig. 9.12b).

224 *Landforms and Geology of Granite Terrains*

9.4 GUTTERS OR RUNNELS

9.4.1 *Terminology*

The flattish, upper surfaces of bornhardts, and the surfaces of platforms or rock pediments are frequently scored by channels cut in fresh granite (Figs 8.18, 8.19, 9.2a and 9.13). These forms are known by various names: Rille, Granitrille, Silikatrille, Karren, Pseudokarren, lapiés, lapiaz, cannelures, acanaladuras and so on. The channels of gentle slopes commonly merge downslope with those of steeper inclines. Here, those on gentle gradients are referred to as runnels or gutters, from their similarity to the roof drains of houses. As on steep slopes, joint-controlled forms are known as Kluftkarren, slots or clefts (Figs 8.18 and 9.6b). Those that are fed principally by seepage from patches of regolith or soil are called decantation runnels or gutters.

9.4.2 *Description*

Gutters several centimetres deep and wide are well-developed in granite in some river beds (Fig. 9.14a). They run parallel to the direction of flow. Scallops and potholes, the latter, due to the grinding action of cobbles and boulders, are commonly associated with them (Fig. 9.14b), and others adopt the shape of a river channel (Fig. 9.14c).

In addition, many flattish granite surfaces are drained by systems of comparatively narrow, deep channels or gutters which link basins to form a rudimentary drainage system (Fig. 9.2a). Most of these gutters are flat-floored and steep-sided. The channel sidewalls are commonly undercut, and some are bordered by raised, levee-like rims (see below). Many of their courses are clearly guided by fractures (Figs 9.6b and 9.13), and some of these are V-shaped in cross section. Such structural influences are not as great, or as consistent, as might be supposed, however. Some gutters, for example, run along joint traces for a few metres and then diverge downslope, following the steepest decline and running across partings with either no discernible diversion or with only minor dislocation. Some drain directly to the margins of the hill where the water runs over steep slopes either in linear channels or in thin sheets spread over the rock surface (and adhering to very steep and overhanging slopes by surface tension). Water draining to the margins of

Figure 9.13. Gutters draining into fracture-controlled Kluftkarren with soil accumulation, Corrobinnie platform, northern Eyre Peninsula, South Australia.

Minor forms developed on gentle slopes 225

platforms disperses in the surrounding regolith. Such systems of connected channels develop only on extensive, gently-sloping surfaces. Where relics of gutters are preserved on comparatively small blocks (Fig. 9.15), it is inferred that the host blocks are remnants of a former dome or platform.

Figure 9.14.   (a) Gutters in granite exposed in channel floor, Ashburton River, northwest of Western Australia. (b) Pot-hole, with grinders (cobbles), in bed of Umgeni River, KwaZulu/Natal. (c) Meanderig gutter on Ampidianambilahy, Andringitra Massif, Madagascar.

226 *Landforms and Geology of Granite Terrains*

Figure 9.15. Relic runnel or gutter segments (a) on granite block near Bruce Rock, southwest of Western Australia, (b) on mushroom rock at Caloote, western Murray Basin, South Australia.

9.4.3 *Origin*

As it has been mentioned, some gutters run along fractures, exploiting the zones of more rapid and intense weathering associated with access of water to the rock. Others, though straight, are not associated with obvious partings, but are probably aligned along zones of stress within which crystals disturbed by strain are especially susceptible to weathering (Fig. 9.16). Slope is, however, a prime determinant of channel orientation and hence pattern.

Minor forms developed on gentle slopes 227

Figure 9.16. (a) Roughly linear depression lacking apparent fractures, Little Wudinna Hill, Eyre Peninsula, South Australia, and (b) nearby Kluftkarren in the same orientation.

That mechanical abrasion is effective in the erosion of the beds and sidewalls of channels is suggested first, by the volume and velocity of water that, charged with mineral particles and other flotsam and jetsam, courses along the channel, if only from time to time; second, by the development of small pot-holes in some localities; third, by the sinuous or meandrine form of some gutters (Fig. 9.14c), which implies lateral erosion of the sidewalls; and last by the basal widening of some bedrock channels, though duration of wetting also plays a part in this development. Channels excavated in granite in periglacial, recently deglaciated, areas may have been eroded by subglacial streams flowing under high hydraulic pressure, with mechanical abrasion both contributing significantly to the end result.

On the other hand, the development of pitted surfaces in channel beds argues the effectiveness of chemical weathering there, and suggests that standing or retained moisture (depending on slope) can cause the gradual loosening and detachment of the rock particles, and their subsequent evacuation by the occasional flows of water. The flask-shaped cross-sections displayed by some channels may be due to longer wetting, and hence more effective weathering of the bed.

As with rock basins, there is clear evidence that some gutters are initiated at the weathering front. At several sites on Eyre Peninsula, South Australia, gutters have been traced into the natural subsurface. On both Dumonte and Crowder rocks, in the Minnipa-Wudinna district, what are separate channels on the exposed low dome converge and coalesce beneath the natural soil level, just as do streams and channels on exposed surfaces (Fig. 9.17). The regolith also is affected by runoff from the nearby rock surfaces. Piping and even sinkholes and elongated subsidence depressions are developed in various granitic areas of the Iberian Peninsula.

The reason for such subsurface extensions of channels is that water pours into the regolith from the domes and platforms but is concentrated in the areas where runnels and gutters debouch from the hills on to the plains. Here the weathering front is lowered relative to the rest of the interface between altered and fresh rock. Shallow channels are formed aligned normal to the edge of the outcrop. Where such piedmont zones have been exposed, pitting is markedly more pronounced in these linear zones than on the rest of the weathering front. Some incipient or subsurface channels

Figure 9.17. Reservoir at Dumonte Rock, near Wudinna, Eyre Peninsula, South Australia, showing gutters draining naturally exposed surface and continuing along exposed weathering front (below X–X). Some even converge in the natural subsurface.

become shallow and wide, as a result of the dispersion of flow between soil fragments, and disappear only one or two metres away from the edge of the outcrop. Others, however, extend several metres at least along the sloping contact between fresh and weathered rock. After exposure the channels are enlarged and otherwise modified, but some of them at least originate at the weathering front and are therefore intrinsically of etch character.

9.5 ROCK LEVEES

At a few sites gutters are bordered by raised rims, standing a few centimetres above the adjacent slope, and known as rock levees, or rimmed gutters (Fig. 9.18a), after their morphological and locational similarity to alluvial forms, which are, however, developed on an altogether larger scale. Such forms occur on Domboshawa (Whitlow and Shakesby, 1988), a granitic bornhardt near Harare, Zimbabwe, where the slopes are scored by broad flat-floored valleys drained by narrow channels or gutters many of which are bordered by rock levees. The rims or levees have been attributed to the precipitation of a protective patina of iron oxides from waters spilling out from the channels, for which there is local evidence around some basins (see below) but which does not occur consistently on the raised rims. They have been attributed to protection by a patina of opaline silica. The absence of lichens (such as *Heppia spp.*), due to moist conditions adjacent to the channels, has also been noted, and it has been suggested that this has a conservative function insofar as on the adjacent flat divides lichens and algae actively weather the granite and render it susceptible to erosion by wash.

An alternative explanation is suggested by observations on Domboshawa. The flattish floors adjacent to the levees are pitted and only sparsely colonised by lichens, suggesting that these areas may once have been covered by a thin regolith (Fig. 9.18b). It is envisaged (Fig. 9.19) that bare rock was exposed in the areas adjacent to the gutters, the edge of the regolith having fallen and been washed into the channels. The exposed rock surfaces were comparatively dry, and hence stable. The regolith, however, retained moisture, so that the weathering of the underlying bedrock continued. It was lowered, leaving the near-channel zones in relief as rims or levees. Siliceous and

Minor forms developed on gentle slopes 229

Figure 9.18. (a) Rock levee in granite and (b) valley with channel, levees and pitted adjacent floor recently exposed from beneath regolithic cover, both at Domboshawa, near Harare, Zimbabwe.

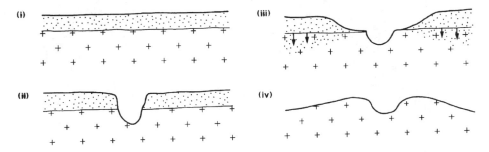

Figure 9.19. Suggested development of rock levee and rock doughnut.

230   *Landforms and Geology of Granite Terrains*

ferruginous patinas could have accumulated at the base of the regolith (see Chapter 3), or near the source of waters carrying the minerals, namely the small streams flowing in the gutters.

9.6 ROCK DOUGHNUTS

9.6.1 *Description*

Rock doughnuts are raised rims surrounding rock basins (Fig. 9.20). They have been described from granitic environments in the Llano of central Texas, and from northwestern Eyre Peninsula, South Australia, and are also developed in Zimbabwe.

Figure 9.20. (a) Rock doughnut in granite, Enchanted Rock, Llano, central Texas. (b) Rock doughnuts in granite near Harare, Zimbabwe.

9.6.2 Origin

Before concluding that he was unable to account for rock doughnuts, Blank (1951) considered a number of possibilities. He wondered whether waters held in the pit had impregnated and indurated the surrounding rock. He could find no supporting evidence from his study area in Texas, and the low porosity and permeability of the crystalline rocks in which doughnuts are developed renders the proposal inherently unlikely.

Blank (1951) speculated whether the flow of water over the surface of the granite domes he studied had been disturbed by the standing water in the basins in such a way as to protect the immediately adjacent rock surface, thus allowing the development of a raised rim. Experimental work and field observations concerned with both wind and water confirm that such obstacles indeed interfere with flow and induce further turbulence, but the effect is to increase, rather than decrease,

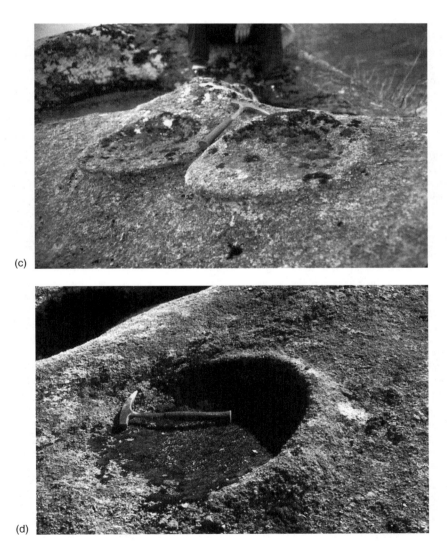

Figure 9.20. (c) Twin fonts on granodiorite from Altar de Cabrões, Serra do Gêrez, northern Portugal. (d) Half doughnut from Río Vilamés, Serra do Xurés, southern Galicia, Spain. The surface on which these forms have developed is more than 300 Ky old.

scouring and erosion. The granite and gneiss outcrops studied are, however, and to a greater or lesser degree, covered by foliose lichen and moss. Though lichens can cause rock disintegration the lichen present on the rims could also conceivably protect the surface, as required by Blank's (1951) working hypothesis.

If the areas marginal to rock basins were favourable to lichen growth because of their dampness, it could be argued that this ring of vegetation could divert flow and wash around the basin, induce increased turbulence and scour, and thus produce the annular rim that is essential to the formation of the doughnuts under investigation. Unfortunately for this suggestion, there is no evidence of pronounced lichen growth on the Texas doughnuts.

Blank (1951) also considered whether the doughnuts could have originated as small, low topographic domes left behind in relief as remnants of circumdenudation by the development of gutters on the sloping granite surface. He visualised basins developing in the crests of these, leaving the resultant rims as rock doughnuts. The initiation of basins in such situations, and the development of doughnuts as a consequence of such local relief inversion, is highly unlikely because water runs off upstanding areas. In special circumstances, such as the fortuitous exposure of a concentration of weak materials on the crest of an upstanding rock mass, such development can be envisaged, but this must surely be a rare occurrence; and doughnuts, covered by lichen, have now been located on several granite residuals on northwestern Eyre Peninsula, South Australia, including some on recent, artificially exposed, platforms.

9.6.3 *Evidence and argument*

Several general arguments can be marshalled against such localised relief inversion giving rise to, and developing rock doughnuts. Yet forms and processes which substantially corroborate this last of Blank's (1951) suggestions have been noted in the field. First, as noted earlier, in several places isolated blocks and boulders rest on plinths, which, though in physical continuity with the underlying granite, nevertheless stand higher than the adjacent slopes. They can evidently form in the subsurface or under the influence of epigene processes, for at both Tcharkuldu Hill and Mt Hall, on western Eyre Peninsula, South Australia, angular blocks protrude above the level of a regolith-covered platform.

Second, boulders standing on plinths commonly develop basal tafoni as a result of moisture attack. Basins are also formed in the plinths (see Chapter 10, and particularly Fig. 10.11). Allowing that the boulder must eventually disintegrate, so exposing the basins set into the plinths, doughnuts could be formed in this way. Thus, the deducible consequences of the working hypothesis involving relief inversion are found in the field. The explanation offered is consistent with the evidence that rock doughnuts are residual forms due to the weathering and erosion of a roughly circular zone inside them, and of an annular ring outside the rim.

Another possible explanation is suggested by the earlier interpretation of rock levees (above). It was suggested that the forms could reflect the contrasted susceptibility to weathering of moist, regolith-covered surface on the one hand and dry, bare surfaces on the other. The same mechanism could account for rock doughnuts (Fig. 9.19). The bedrock areas exposed around basins are relatively dry and stable, whereas the bedrock beneath adjacent bedrock-covered platforms and slopes is weathered more rapidly, leaving the circular area around the basin in relief. This hypothesis finds support in the occurrence of two small doughnuts on a recently cleared rock platform at Kwaterski Rocks, north of Minnipa, Eyre Peninsula, South Australia.

9.7 FONTS

In some granitic areas in Portugal, Galicia and Catalonia, Spain (Vidal Romaní and Twidale, 1998; Roqué and Pallí, 1991), rock doughnuts and basins occur on the crests of pedestals or small towers, together making fonts or benitiers which are commonly up to one metre, but in the Sierra Guadarrama of central Spain (Centeno, 1988), up to 4 metres high (Fig. 9.21). The forms are due

Figure 9.21. Font, some 4 m high, in granite, Sierra Guadarrama, central Spain (Centeno, 1988).

to preferential etching and can be explained in terms of weathering and erosion of the areas adjacent to basins. As the surface around the doughnuts is lowered, they develop fonts when the diameter of the crestal basin is less than its height above the adjacent platform. Every stage in this process of lowering of the surface surrounding basins can be seen at some localities in the Serra do Gêrez (Coudé Gaussen, 1981) and the Sierra de Guadarrama, both on the Iberian Peninsula.

REFERENCES

Blank, H.R. 1951. "Rock doughnuts", a product of granite weathering. American Journal of Science 249, 822–829.
Boyé, M. and Fritsch, P. 1973. Dégagement artificiel d'un dôme crystallin au Sud-Cameroun. *Travaux et Documents de Géographie Tropicale* 8: 69–94.
Centeno, J.D. 1988. Morfología Granítica de un Sector del Guadarrama Occidental (Sistema Central Español). Editorial, Universidad Complutense de Madrid.

Coudé-Gaussen, G. 1981. Les Serras da Peneda et do Gêrez. Etude Géomorphologique. Universidade de Lisboa. Instituto Nacional de Investigação Científica. Memórias do Centro de Estudos Geográficos, Lisbon.

Fry, E.J. 1926. The mechanical action of crustaceous lichens on substrata of shale, schist, gneiss, limestone and obsidian. *Annals of Botany* 41: 437–460.

Hedges, J. 1969. Karst caves in silicate rocks. D.C. *Speleo* 3–4.

Jones, T.R. 1859. Notes on some granite tors. *Geologist* 2: 301–312.

Lageat, Y., Sellier, D. and Twidale, C.R. 1993. Mégalithes et méteoritisation des granites en Bretagne littorale (France du Nord-ouest). *Géographie Quaternaire et Physique* 48: 107–113.

Roqué, C. and Pallí, L. 1991. Modelat del Massis de Begur. Estudis Sobre el Baix Empordà 10: 5–44.

Scott, G.D. 1967. Studies of the lichen symbiosis. 3. The water relations of lichens on granite kopjes in central Africa. *The Lichenologist* 3: 368–385.

Smith, L.L. 1941. Weather pits in granite of the southern Piedmont. *Journal of Geomorphology* 4: 117–127.

Twidale, C.R. and Bourne, J.A. 1975. The subsurface initiation of some minor granite landforms. *Journal of the Geological Society of Australia* 22: 477–484.

Twidale, C.R. and Campbell, E.M. 1992. On the origin of pedestal rocks. Zeitschrift für Geomorphologie 36: 1–13.

Twidale, C.R. and Corbin, E.M. 1963. Gnammas. *Revue de Géomorphologie Dynamique* 14: 1–20.

Twidale, C.R. 1988. Granite Landscapes. In Moon B.P. and Dardis G.F. (Eds), The Geomorphology of Southern Africa. Southern Book Publishers, Johannesburg. pp. 198–230.

Vidal Romaní, J.R. 1989. Granite geomorphology in Galicia (NW España). *Cuadernos Laboratorio Xeolóxico de Laxe* 13: 89–163.

Vidal Romaní, J.R. and Twidale, C.R., 1998. Formas y Paisajes Graníticos. Serie Monografías 55. Universidade da Coruña. Servicio de Publicacións, A Coruña.

Watson, A. and Pye, K. 1985. Pseudokarstic micro-relief and other weathering features on the Mswati Granite (Swaziland). *Zeitschrift für Geomorphologie* Supplement Band 29: 285–300.

Whitlow, J.R. and Shakesby, R.A. 1988. Bornhardt micro-geomorphology: form and origin of micro-valleys and rimmed gutters, Domboshava, Zimbabwe. *Zeitschrift für Geomorphologie* 32: 179–194.

Worth, R.H. 1953. Worth's Dartmoor. In Spooner G.M. and Russell R.S. (Eds). David and Charles, Newton Abbott.

10

Caves and tafoni

10.1 GENERAL STATEMENT

Caves are underground openings connected with the atmosphere, in places by wide openings, elsewhere by narrow apertures or vertical shafts, but at some sites only by narrow gaps in the surrounding rock. Granite masses are not noted for cave developments, but several have been reported from various parts of the world. On the other hand, tafoni, or relatively shallow inverse or laterally disposed hollows, are widely developed on granitic rocks.

The term pseudokarst has been applied to forms which mimic features developed by solution in carbonate or gypsiferous strata but which are developed in rocks not as obviously or rapidly soluble. Thus, various so-called pseudokarstic forms, including caves, are developed in acid plutonic rocks in several parts of the world. Used in this context, the prefix pseudo is disputed by those who point out that solution contributes significantly to the formation of flutings, gutters, basins and caves in granitic rocks (see also Chapters 8 and 9). Others, while not denying the efficacy of solution as a major weathering process acting on all minerals, including all the essential minerals of granite, nevertheless argue that other mechanisms such as abrasion are more important here than in carbonate rocks. In any event, an alternative term such as silicate karst (hence Silikatrillen, Granitrillen) might be more appropriate (Kastning, 1976).

10.2 CAVES ASSOCIATED WITH CORESTONES AND GRUS

Occasionally, the preferential subsurface flushing or evacuation of friable, weathered rock has left irregular, tubular voids, largely defined by corestones set in grus and known as boulder caves. Such subsurface erosion is achieved by streams that are diverted underground and re-emerge a short distance downstream.

Thus, Labertouche Cave, near Neerim South, in Victoria, is about 200 m long and essentially straight, though irregular in detail. It begins in a sinkhole in a blind valley and its point of emergence is marked by a pronounced re-entrant. Others of a similar kind have been reported from the uplands of eastern Victoria, in the Girraween National Park, New South Wales, and from central and southeastern Queensland. They are also developed in Spanish granite massifs in the Pyrenees, and in Galicia (in the Pindo, Louro-Os Profundos and Barbanza uplands) where the well-fractured rock has been, subject to Holocene periglacial activity. Cave systems in granite are also reported from various parts of the Western Cordillera of the USA.

Granite caves present the same hazards as karst caves. In 1980 a caver died from injuries received when he fell and suffered immersion in a boulder cave, one of the Lost Creek Granite Caves, in Colorado, which, though up to 15 m high, are also characterised by boulder falls and by streams that from time to time run at high velocity.

A cave system is developed on the granitic Makatau inselberg in the Rupununi savannas of Guyana, where the form of the openings in plan is clearly influenced by the orthogonal joint system.

236 *Landforms and Geology of Granite Terrains*

The caves run for some 60–70 m between corestones. These caves appear to be due to the piping and flushing of grus. The corestones and grus either remain sufficiently rigid to form a roof and sidewalls, or the corestones subside toward the evacuated area, but are packed closely enough to jam together and form a roof of reasonable stability.

Again, some ten granite caves, varying in length between two and fifteen metres, have been reported from the High Tatra Mountains of Poland. Some appear to be of the same origin as those already considered, but others are attributed to preferential solution acting along sideritic veins. There are also vertical shafts and niches in the Karkonosze Mountains, also in Poland, where they are associated with feldspar-rich zones. Calcite speleothems are developed on the ceilings of some of these caves, and also in caves in the Spanish Pyrenees, in Lleida Province (Vidal Romaní and Vilaplana, 1984), or in Andringitra Massif, Madagascar, where the dissolution of plagioclase provide calcium (Vidal Romaní, Twidale, Bourne, 1998).

10.3 CAVES ASSOCIATED WITH FRACTURES

Weathering along sheeting planes has in many places created gaping partings. At a few sites such weathering has proceeded to such an extent that the openings are large enough to be called caves. Thus, the Enchanted Rock caves, located on the bornhardt of the same name in the Llano of central Texas, are developed at two levels (Fig. 10.1). The upper cave is some 250 m long, the lower about

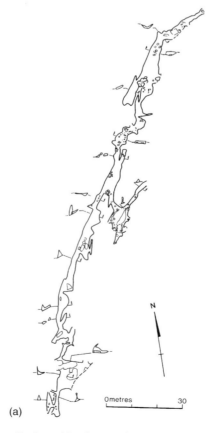

Figure 10.1. (a) Plan of caves at Enchanted Rock, central Texas.

70 m. Each is 2–3 m high. Both follow along major sheeting joints and appear to be due to the weathering of the granite by moisture, with the greater part of the cave development taking place on the underside of the upper sheets (Fig. 10.2).

Boone's Cave, in North Carolina, USA, is similar. Developed in a granite-gneiss, it is some 40 m long, displays obvious joint control in plan, and is thought to be essentially an enlarged bedding plane. Similar fracture-controlled caves in granitic rocks, and totalling some hundreds of metres in length, are developed in California, and an opening, some 35 m long and aligned parallel with the hillslope and a sheet fracture, is reported from near Seoul, in South Korea.

Figure 10.1. (b) Cave formed along sheet fracture, Enchanted Rock, central Texas.

Figure 10.2. (a) Tafone developed on a parallelipipedic block of granodiorite in Monte Louro, Galicia, NW Spain.

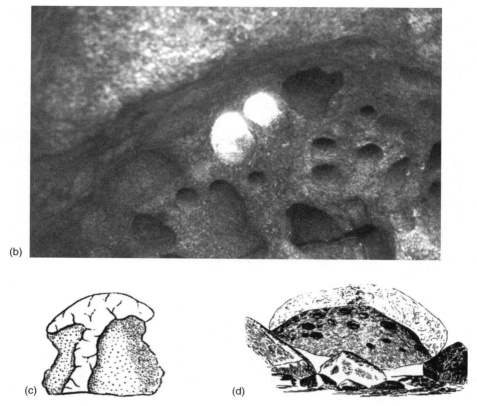

Figure 10.2. (b) Tubular aperture in the outer shell of a tafone from Mount Pindo, Galicia, northwestern Spain. (c) Tafone sketch from Guadarrama, Madrid, Spain (Casiano de Prado, 1864). (d) Tafone sketch from Galicia, Spain (Hult, 1888).

10.4 TAFONI

10.4.1 *Description*

Tafone (plural tafoni) is an Italian word meaning, variously, depending on regional usage, a perforation or a window. As applied in geomorphology it means a shallow cavern or a hollow partly, and frequently substantially, enclosed through the preservation of a visor, or hood. Some tafoni are small (alveoles -see below- a few centimetres wide and high), but some are several metres wide and of a similar height. The walls of the hollows may be regular, but others are flaked and mamillated, and some, as mentioned previously, display shallow flutings. Some tafoni are connected to the outside by breaches or windows in the outer shell of rock (Fig. 10.2a). Some such apertures are tubular (Fig. 10.2b), but others are quite wide and of irregular shape. The partial enclosure of the hollow differentiates it from an alcove (e.g. Fig. 3.7). Large alcoves are called grottes, or, in Australia, shelters, though these are usually located at the base of cliffs and are called cliff-foot caves.

Tafoni were first described by De Prado (1864) from the Sierra de Guadarrama in central Spain (Vidal Romaní, 1998 and see fig. 10.2c) and later by Reusch (1883) from Corsica, and by Hult (1888) in Galicia NW, Spain (Fig. 10.2d), but normally the most known citations are the one of Penck (1894), of forms he had observed in Corsica (Klaer, 1956). Tafoni are especially well-developed in granite, where they are commonly found at the margins of sheet structure, beneath boulders and in the scarp-foot zones of bornhardts (Fig. 10.3). Alcoves or shelters lacking visors

Caves and tafoni 239

Figure 10.3. Tafoni developed (a) along sheeting plane (b) beneath boulder, both at Ucontitchie Hill, Eyre Peninsula, South Australia. (c) At the scarp-foot, merging laterally with flared slopes, at Kokerbin Hill, southern Yilgarn of Western Australia.

Figure 10.4. (a) Open shelter or alcove occupying void left by evacuation of joint-defined block, at base of granite tor, or castle koppie, on Dartmoor, southwestern England (Geological Survey Museum, UK). (b) Poorly developed visor on shelter at Mt Hall, northwestern Eyre Peninsula, South Australia.

and due to basal sapping, to the exploitation of intersecting fractures (or the removal of the joint blocks so defined), are also widely developed (e.g. Fig. 10.4).

Very small hollows are called alveoles, hence alveolar weathering (Mustoe, 1982). Alveoles can form quickly, for some have already developed on a seawall constructed of greywacke (sandstone) on the Victorian coast in 1943.

When alveoles are associated in numerous groups on the interior walls of tafoni, many authors refer to them as honeycomb weathering (Fig. 10.5). The disappearance of this honeycomb structure in the very well evolved tafoni and the substitution of it by a scalloped surface (Fig. 10.6b) suggest that honeycomb development is an early stage of tafoni evolution process. Generally, the

Caves and tafoni 241

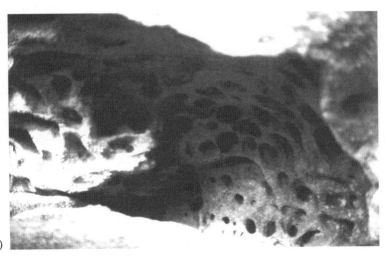

Figure 10.5. (a) Early stage of honeycomb development in a tafone from Girona, Costa Brava, Catalunya, northeastern Spain. (b) Large honeycomb on interior wall of tafone, Lézaro, North side of Río Xallas, Galicia, northwestern Spain.

242   *Landforms and Geology of Granite Terrains*

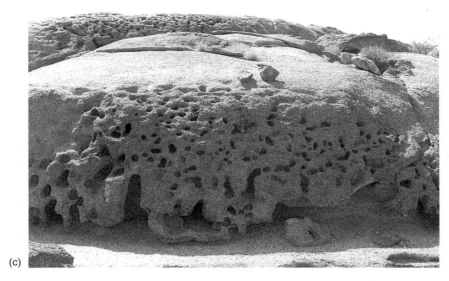

Figure 10.5.   (c) Examples of alveolar weathering at various scales: small alveoles, Zebra Mountains, central Namibia.

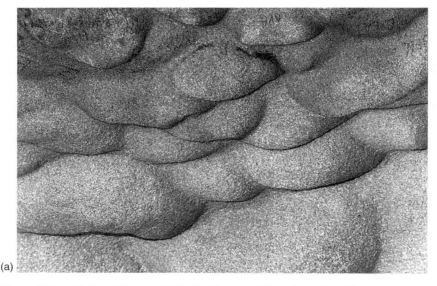

Figure 10.6.   (a) Mamillation at Remarkable Rocks, Kangaroo Island, South Australia.

alveoles found on the interior walls and ceilings of tafoni are larger (10–15 cm diameter) than those preserved on exposed surfaces (2–3 cm diameter). This may be a function of the protected environment within tafoni, where the hollows can extend, as compared with exposed sites where alveoles are destroyed before there is time for them to grow. But also, it is evident that the sizes of the alveoles are related to the grain size of the rock: fine grained rocks give small alveoles while coarse grained rocks give big alveoles. The development of convex sheet surfaces produces a ribbed and mamillated/scalloping effect which is another characteristic internal morphology of

Figure 10.6. (b) Scalloping at The Granites, Mount Magnet, Western Australia. (c) Books of flakes in mamillated ceiling of tafone, Ucontitchie Hill, Eyre Peninsula, South Australia.

tafoni (Figs 10.6a and b). Books of flakes developed over the entire rock surface are commonly associated with such features (Fig. 10.6c). Such flaking denotes an active surface. Elsewhere the granite is loose and disaggregated. But some walls are stable and apparently no longer subject to change.

Tafoni are developed on the undersides of sheet structure and of boulders, though they also occur on the sides of steep rock walls, hence, sheet tafoni, boulder tafoni (tortoiseshell rocks) and sidewall tafoni. Tafoni vary in size from a few centimetres radius to large hollows that are metres across and high, and in which a group of people can readily stand. They have been described from several climatic contexts, and in particular from arid or semi-arid areas, both cold and warm, both interior and coastal, for instance from Antarctica, Hong Kong, Corsica (Klaer, 1956), Sardinia, central and southern Australia (Fig. 10.7). The split mushroom rock known as the Peyro Clabado in the Sidobre of southern France is hollow on its underside, and small tafoni and alveoles, known

Figure 10.7. Examples of tafoni (a) in Antarctica and (b) in Hong Kong, with alveoles and ribs within the tafone (Tschang, 1961).

Figure 10.7. (c) In Corsica with scallops.

locally as dew holes, are reported from the Bohemian Massif. But they do not appear to be developed on the Dartmoor granite, southwestern England, for example.

At a local scale, the distribution of tafoni is puzzling with some blocks hollow and others, immediately adjacent and apparently identical, intact. Such contrasted developments are found, for example, in exposures of the Begur granite, in the Costero-Catalana Range of northeastern Spain. In the Pindo granitic massif of Galicia, tafoni are scarce, though those that are present are very well-developed. Again, although tafoni are widespread in the Upper South East and Eyre Peninsula regions of South Australia, not all boulders exhibit tafoni.

10.4.2 *Process*

The process responsible for tafoni can be considered under three headings: initiation, growth or development, and the nature of the visor/outer shell.

10.4.2.1 *Initiation*
There is no doubt that some tafoni are initiated by soil moisture attack beneath the land surface. Boyé and Fritsch (1973) reported hollows in the fresh rock surface occupied by grus and already present on the underside of sheet structure in a newly exposed quarry face at Ebaka, in southern Cameroon. Similarly, the outlines of tafoni were noted already developed during the excavation of the foundations of a dam at Xallas River, at Ézaro, near A Coruña, in Galicia NW Spain. By its very nature, such direct evidence of subsurface initiation is rare.

In addition, tafoni and flared slopes commonly merge. Examples have been noted in the sidewalls of a joint cleft on Scholz Rock and on boulders at Murphys Haystacks, both on northwestern Eyre Peninsula, South Australia, at Kokerbin Hill, in the southwest of Western Australia (Figs 10.3c), and on Bloedkoppie Dome in central Namibia. Pronounced hollows are associated with basally fretted slopes at several sites on Eyre Peninsula (Figs 8.11a and d), and in the Pietersburg area of the Northern Transvaal (Fig. 10.5b).

Alternatively, many tafoni appear to originate at the land surface. Some workers have attributed tafoni to preferential weathering under epigene, or subaerial, conditions. They argue that the hollows were originally occupied by material that was different from, and presumably weaker than, that which remains. At most sites it is difficult to test this hypothesis because the alleged different material has been eroded, so that comparisons are impossible. On the other hand, it is clear that granites, like other igneous and metamorphic crystalline rocks, display significant primary petrological variations, both textural and compositional, and patches of such mafic minerals as biotite may well have been weathered more rapidly than the mass of rock, producing a small initial depression or hollow. For instance, on the Costa Brava, in Girona, northeastern Spain (Fig. 10.5a), an example has been noted of a tafone developed in a block the interior of which appears to be altered and deformed, in contrast to the outer zones. Small alveoles are developed in the deformed rock. Alveoles undoubtedly coalesce, but such a mechanism can scarcely account either for the enormous tafoni observed in the field, for their preferred development on the undersides of blocks and sheet structure, or for their common occurrence in the scarp-foot zone. The common experience is that tafoni develop without regard to compositional or textural variations in a rock mass.

Other investigators have suggested that concavities developed in the rock surface subsequently evolve into tafoni. The concavities are said to be either structural in origin and related, for example, to curved joints, or to result from uneven scaling. Again, however, the suggestion provides no explanation of the development of the forms or for their distribution.

Vidal Romaní (1983) suggests that high load concentration at few points of the block/base causes the plastification of the rock in volumes of varying dimensions (lacunar zones). This phenomenon may be produced either in subedaphic environments (Figs 10.8 a and b), or in tectonically strained areas (Fig.10.8c). Inside these zones the rock will be with greater susceptibility to alteration and so more susceptible to moisture attack. The formation of lacunar domains is carried out by deformation of the rock in its solid state and can be contemporary with (tectonic process) or later than (edaphic process) the definition of discontinuities (diaclases) which usually affect the rock. Afterwards the epigenic alteration, either subaerial or subedaphic, displays them up as tafoni or gnammas when the rock held in them alters at a faster rate than the rest of the rock. Similar features were described as formed below the ice in glacial environments by Drewry (1986).

10.4.2.2 *Growth or development*
The development of tafoni has been attributed to various processes and mechanisms. Temperature variations within tafoni have been cited as sufficient to cause disintegration of the rock exposed in the walls and ceilings of the hollows and so bring about the enlargement of the negative forms. Others have argued that the microclimates found in tafoni are more stable with a smaller range of temperature and humidity than is experienced in the open air. It has been suggested that hydration is more marked on the interior surfaces of tafoni than on the external walls of the rock masses, and that the minerals so affected cause fracturing and flaking (the negative exfoliation of some authors) of the rock. Granular disintegration is attributed by some to similar processes. Unfortunately, the rock forming the flakes and granules is essentially fresh and frequently displays little or no sign of alteration.

Ikeda (1990, 1994) suggests that some tafoni in Korea appear to be enlarged through freeze-thaw activity. Thus, after this author moisture blown into existing openings on the wind freezes in the low winter temperatures causing the surface layers of rock exposed in the ceilings of tafoni to shatter. Ikeda (1990, 1994) interprets the rock meal accumulation on the floor of tafoni during winter, but not in summer as the proof of it. And that the process takes place quite rapidly – on the scale of years. But many active tafoni are in many areas where frosts are few and not severe, and even in Corsica, for example, Klaer (1956) exclude tafoni from areas subjected to alternations of freezing and thawing. Also, how moisture penetrates the rock from the walls and ceilings of tafoni that are essentially enclosed (as is the case in some boulder tafoni) remains unexplained.

Many workers attribute the development of tafoni to flaking or granular disintegration due to haloclasty or salt crystallisation (most commonly halite but including such salts as mirabilite). This is consistent with their common occurrence in arid or seasonally arid lands and on adjacent coasts. Theoretically, salts can disrupt a rock by crystal growth, by hydration expansion, by thermal expansion and by osmotic pressure. Though thermal expansion is favoured by several workers, it has been shown that temperature variations within tafoni are very limited. Hydration expansion, even in conditions of suitable temperature and relative humidity, is rejected as acting too slowly to be

Caves and tafoni 247

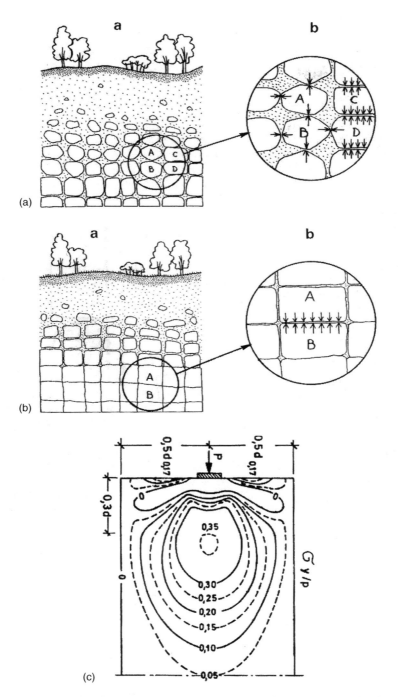

Figure 10.8. Sketch of subedaphical weathering of a granite massif through the discontinuities system, (a) with uniform distribution of lithostatic charges and (b) with concentrated distribution of lithostatic charges (Vidal Romaní, 1989). (c) Sketch of lacunar zone under a concentrated P charge. The domain included in the continuous line 0 corresponds to the area located at compression. The rest of the rock, between 0.05 and 0.35, corresponds to an area that would eventually evolve towards a tafone or a gnamma (Vidal Romaní, 1984).

effective. Osmotic pressure effects require some form of suitable membrane to be effective, and an essentially closed system, neither of which conditions is found in the field. Thus, crystal growth of salts, especially gypsum and halite, remains as the one of the most cited mechanism. It was described more than a century ago:

> "... the tendency of crystals to increase in size when in contact with a liquid tending to deposit the same crystalline substance must push out of their way the porous walls of the cavities in which they are contained..." (Thomson, 1863, p. 35).

Many workers have in various degrees favoured the salt crystallisation mechanism. Some consider that salt contributes both mechanically and chemically to rock disintegration, acting through expansion or crystallisation, and through the ionising action of slowly evaporating water. Others argue that insolation causes salts to expand more rapidly than the host rock – and so on; there are many variations on the salt theme, including some who assign to it merely a contributory or secondary role.

The effectiveness of salt crystallisation in rupturing rocks has been demonstrated in laboratory conditions and experimental work, plus observations of building stones that have crumbled as a result of coming into contact with natural or industrial solutions, have together convinced many researchers of the efficacy of salt crystallisation as a mechanism of rock weathering. Many have concerned themselves with systematic studies of building stones, the disintegration of which is constantly in the public consciousness due to the reported weakening of the fabric of such well-known structures as the Taj Mahal and the Parthenon as a result of industrial air pollution.

Salt crystallisation has long found favour with field geologists and geomorphologists (Bradley, Hutton and Twidale, 1978). Many years ago Jutson (1917) referred to exsudation; various German workers have described Salzsprengung, and Klaer (1956) has strongly argued the case for salt-induced disintegration in Corsica. For obvious reasons, the process has been invoked in explanation of weathering in general, and of hollows called alveoles or tafoni in particular, in coastal and arid environments, for it is there that salts are most readily available. Many have invoked salt crystallisation in explanation of honeycomb and similar weathering forms in coastal contexts, while several have resorted to the process in hot deserts. Salt crystallisation in polar regions, particularly Antarctica, has been cited by many writers.

The experimental and field evidences have together convinced many that salt crystallisation is responsible for the enlargement by means of some tafoni. But, serious problems such as the cause of mamillation, scalloping or honeycomb remain as well as the generalized absence of salts in the grain accumulations produced by disaggregation on the tafoni base. The origin or source of the salts said to have crystallised and caused rocks to rupture has been, and remains, a matter of concern. The association of some tafoni exposures with coastal environments has encouraged the suggestion that the salts are of marine origin, and that they are carried on to the rocks either by waves or in fogs. Thus, in western Namibia both alveoles and tafoni are attributed to uneven granular disintegration due to salt crystallisation, the salt having been introduced in sea fogs that are common there. Other workers invoke the crystallisation of salts derived directly from the sea. Some regard the salts as cyclic or transported by the wind; hence, it is said, the profuse development of tafoni in deserts, both hot and cold. Salt is certainly transported in this manner. Clouds of salt have been seen being blown from the surfaces of salinas by strong winds in several parts of Australia. The difficulty, both with marine and aeolian salts, or indeed salts carried in rivers or groundwater, is to explain how they can penetrate deep into rocks of low permeability. But given time, they could infiltrate in solution along fractures and along associated stress fissures. Hence, those many tafoni associated with joints.

Connate salts, or salts derived from the rock itself, present no such problem. Most granites contain radicals of Ca, Na, SO_4 and Cl in their feldspars and micas. These are released on weathering, and combine and crystallise as halite or gypsum, exerting enough pressure to rupture thin particles of rock as they do so. In order for the salts to be translocated within the weathered outer shell of the rock, however, water is needed. On the exposed upper surface and sides of boulders, for example, leaching of salts by rain is likely. Moreover, on such exposed surfaces, lichens and mosses grow and absorb water. But, it is on sheltered undersurfaces that tafoni develop for the most part. Here, dew may form, or the vapour respired by small mammals that find shelter in the crevices and hollows may be condensed.

The question is however more complex. For instance, chlorine is present in several British granites (Thomson, 1863), and of the areas examined chlorine attains its highest concentration in the granites of southwestern England. Yet, neither there nor elsewhere in Britain are true tafoni in evidence. Possibly in humid climates salts are washed through the system before they can cause rock disintegration. It is also difficult to understand how salt in concentration no greater than 0.5% can create high porosity in otherwise fresh rock and cause it to rupture.

10.4.2.3 *The nature of visor/outer shell*
By definition, it is critical to the formation of tafoni, for it differentiates enclosed tafoni from open alcoves and shelters, but it is difficult to explain. Flaking and disintegration wear away and eventually breach the pendant slab from the inside, suggesting that whatever has protected the visor/outer shell is to be found on the outer surface. For Wilhelmy (1964) the visor/outer shell merely reflects an outer skin of rock that has been dried and thus hardened. Certainly, thin surficial zones, lacking any discernible alteration, but nevertheless clearly more resistant than those immediately below them, have been noted in the field. Alternatively, there are many examples, and not only in granitic rocks, of secondary mineralisation having taken place along fractures, with salts having precipitated out from circulating groundwaters. Such impregnations could account for some hardened outer zones of blocks and hence for visors. But the outer casing is breached more and more and, and it is eventually eliminated, so that the distinction between tafoni and alcoves is in granite terrains evolutionary (see below); though elsewhere, as in limestone and sandstone, alcoves are genetically different from their granitic counterparts. In other cases, especially areas of wet climate, the external zone of the block is covered by a mostly continuous layer of lichens.

10.4.3 *Stages of development*
The vast majority of tafoni starts from fracture planes whatever their orientation (Fig. 10.8). Some few are apparently initiated on weathered or eroded surfaces such as flared slopes, possibly on xenoliths of more readily weathered mafic minerals, but a fracture plane is a much more common starting point. The first stage (i) starts when hollows form on both sides of the fracture plane (Fig. 8.12b). In general, however, the hollow in the upper block is larger than that in the lower (Fig. 10.9), the reasons of these differences are evident when the starting surface is horizontal. Weathered bedrock, due either to haloclastic flaking, granular disintegration or other, falls on to

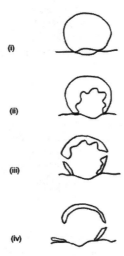

Figure 10.9. Suggested stages in development of tafone.

Figure 10.10. Development of hollows on opposed upper and lower surfaces of a basal parting: Tcharkulda Hill, near Minnipa, northwestern Eyre Peninsula, South Australia.

the fracture plane, and accumulates there in thin discontinuous covers of fragmented rock until washed or blown away by wind or rain. For the case of surfaces of vertical or strongly inclined starting surfaces, the development of both cavities is even. In the second or anisotropic stage (ii), the active surface advances unequally, rapidly in alveoles and mamillated surfaces, less so in the septa and ribs in between. In places the intervening rises form orthogonal patterns. Granular disintegration is still active, but thick books of flakes are spectacularly developed at some sites. The hollow is thus enlarged. The dryness of the interior walls is attested by the absence of lichens. Swifts also build nests, and similarly, bats inhabit the hollows and their droppings, in places attain considerable thickness, may contribute to the formation of speleothems. The third stage (iii) corresponds to the moment in which overall, the rate of development of the tafoni decreases, though alveoles continue to extend, causing the intervening septa to disappear.

Eventually, the extending hollow intersects the outer surface of the host block or boulder. The apertures or windows thus formed differ in their morphology. In some, the outer and inner surfaces of the confining shell are roughly parallel and the window takes the form of a bevelled opening. Elsewhere the aperture in section looks like a truncated cone (Fig. 10.2b), narrowing toward the outside; they have been referred to as ox-eyes.

The fourth stage (iv) is characterised by the breaching of the outer shell that causes marked changes in the microclimate in the tafoni. At this stage the interior ceilings are scalloped. Eventually in this stage, the water entrance from the exterior may take place, so that small siliceous speleothems derived from the dissolution and reprecipitation of silica also appear (see below). But because of increased air circulation and accession of moisture consequent on the opening of the hollows, the walls are colonised by biota such as algae, lichens and mosses; and these may accelerate extension due to the general breaching of the confining shell and the collapse of the remains of the block or boulder. All that remains is a scatter of fragments and detritus.

10.4.4 Case-hardening and other veneers

Case-hardening is commonly invoked in explanation of the visors/outer shell enclosing tafoni. It takes the form of a red-brown coating concentrated at the surface, but extending several crystals deep beneath it that is commonly developed on exposed surfaces in arid and semi-arid lands. It is associated with projecting lips, is obviously more resistant than the unaltered rock, and, for this reason, is widely referred to as case-hardening. Although some writers use the term almost synonymously, case-hardening differs from desert varnish, which term is used of a wide range of black, brown, yellow-brown and colourless patinas found in a wide range of climatic conditions, but particularly in the arid and semi-arid tropics.

The brown, red, yellow and purple patinas are composed of silica and oxides of iron and manganese in various proportions. Case-hardening stands in contrast not only with true varnish but also with the black coatings frequently found in association with water (pools, rivers) in various climatic environments, for this reason, known as river films. They are supposed to be of organic origin, and to consist of the remains of algae and lichen that have been concentrated in streams and wash.

The black encrustations so common in northern Australia and southern Africa, for example, may be of similar origin in part, though soot derived from seasonal (anthropogenic) burning may also contribute. Even so, the coatings are especially thick in gutters that score the bare rock surfaces. On Eyre Peninsula, South Australia, several of the gutters draining Wudinna Hill and Ucontitchie Hill carry encrustations of black material, and on Yarwondutta Rock and Turtle Rock some such grooves extend on to the overhanging slopes of the flared margins of the residuals. The black veneer is thin, discontinuous and relatively weak. It can, for example, be scraped away by sharp, needle-like leaves to give scratch circles. But, at Yarwondutta Rock and elsewhere the coating is associated to the most stable parts of the rocky surface disappearing the most unstable ones (Chapter 8). Turning to the origin of case-hardening, many of the comments made in respect of varnish apply equally to case-hardening. Engel and Sharp (1958), however, demonstrated that rust-coloured case-hardening developed in the arid American Southwest consists of oxides of iron, manganese, silica and alumina. Electron probe work shows that, though the patina is uniformly thin (it is nowhere more than 100 microns thick and mostly less), it consists of two layers: an inner coat consisting of SiO_2 and Al_2O_3, but with some iron and manganese, and an outer zone composed wholly of oxides of these last named metals. Similar studies carried out on varnish developed on an olivine basalt from Arizona have demonstrated a layered and botryoidal structure, but, again, a dominance of iron and manganese. The same study distinguished optically opaque or dark layers rich in manganese and lime, and red layers depleted in manganese but rich in oxides or iron, alumina, silica and potassium.

Some writers have suggested a biological origin for the varnish of arid lands. Scheffer, Meyer and Kalk (1963), for instance, regard it as due to blue algae which have oxidised iron and other heavy metal ions, and concentrated them in superficial oxidised skins on stones and other surfaces. Dorn and Oberlander (1981) attribute desert varnish to *Metallagenium*-like bacteria capable of concentrating ambient manganese, silicon and iron on rock surfaces.

Another possible explanation has been prompted by the observation that iron oxides (probably goethite and haematite) are concentrated at the weathering front in some profiles developed on granitic rocks. Some are very pronounced, some only faintly visible, but such concentrations are commonly present. Below, the rock is intrinsically fresh and cohesive; above, the thin iron oxide zone is bleached and altered, presumably as a result of the leaching and illuviation of soluble salts from the surface to the weathering front, where they are concentrated, the solutions being able to penetrate no further into the impermeable rock. This bleached, outer zone, which varies in thickness between 2–3 mm to 4–5 cm, remains hard when dry, but becomes soft and friable when wet.

It is suggested that granite masses, whether boulders or inselbergs, initiated in the subsurface by differential weathering, may acquire such a marginal concentration of iron oxides; that the weathered, outer, bleached zone is eroded after exposure; and that in this way the fresh rock masses come to have a coating or patina rich in iron which is enough to protect the underlying rock against epigene weathering and thus allow a visor to develop.

10.5 SPELEOTHEMS

Weathering inevitably leads to solution (Meybeck, 1987; Bennett, 1991), and the manifold erosional features resulting from the degradation of granitic rocks have their corollaries in various

252   *Landforms and Geology of Granite Terrains*

depositional or constructional forms (Vidal Romaní and Twidale, 1998). The most noticeable one, due to being unknown up to now, are the small siliceous speleothems widely developed in partings and other opened discontinuities of the granitic massifs (Fig. 10.11a). They are here called, as their congeners in soluble rocks, speleothems though they are different as to size and appearance frequency to the ones in the karstic systems. Their forms are very similar, and thus in the literature they have, as far as possible, the same names when the form is the same. There may be distinguished two types of deposits. The first type is the cylindrical speleothems: appear associated to some water dripping points of the fissure system. Within this type of speleothems two subtypes may be differentiated: stalactites, which are formed in the roof of rock fissures in the dripping point of the water that circulates through the fissure. False stalagmites or antigravitational speleothems: are due to antigravitational growths on the base of the cavities. Its lineal development is limited. The second type of speleothems is developed as layers or coatings and is produced by trickling of water through the rock fissure system. They are called flowstone, dripstone, gourdams or rimstone as their congeners in limestone (Finlayson and Webb, 1985). The elemental

Figure 10.11.   (a) Siliceous speleothems from Kokerbin Hill, Western Australia. (b) Siliceous speleothems from Tcharkuldu Hill, Southern Australia.

analyses carried out on the body of the speleothems indicate a constant composition in Si for all the samples studied independently from their host rock or environment of formation. Moreover, other chemical elements appear in very low proportion such as Al, P, K, Fe, etc., though caolinite speleothems have been mentioned, it is usual that mineralogically the speleothems are of opal-A (Vidal Romaní, Twidale, Bourne, 2003 and see Figs 10.11b and c). The observations under the SEM show an identical morphology for all the samples independently from the climate under which they were developed and from the kind of rock with which they are associated. The starting situation is the dissolution in the water contained in the fissural system of the granitic rocky massif of the Si mainly, and of the other chemical elements forming the rock. This process is carried out by the attack of different bacterial organisms. After this, it takes place the formation of the speleothems by precipitation of the biogenic opal formed by this process. Three phases have been distinguished in the speleothems development. In the first phase (biogenic phase), the texture of the speleothems is porous, brecciate (opal-A clasts) or conglomeratic (opal-A oolites) the first ones (Fig. 10.12a) formed by mechanical flaking when the silica gel dehydrates, and the second ones caused by direct precipitation of amorphous silica by minor organisms (bacteria and fungi). In the second phase (re-solution phase), the high solubility of the biogenic opal may produce the progressive internal silting of the porous system of the speleothem that finally is totally infilled with the eventual formation of external coatings that covers externally the speleothem (patina) (Fig. 10.12b), transforming the final end of it into a solid body with rhytmical texture of accretion equivalent to that of calcareous speleothems (Fig. 10.12c). This wet and rich in silica environment allows a brief existence of different microorganisms (fungi, diatoms, bacteria, etc.), whose life is directly controlled by the water reserves (derived from rain, percolation, etc.). The water absence interrupts the biological activity causing the burial and fossilization of living organisms by

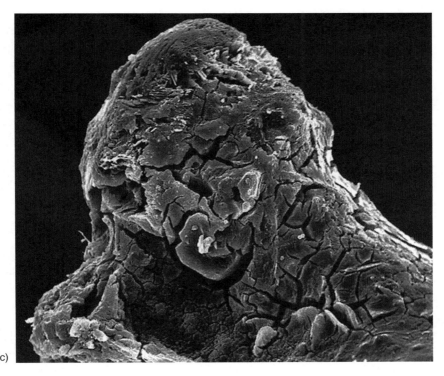

Figure 10.11. (c) SEM image of siliceous speleothem showing the typical cracks of the dessicated opal A, Monte Louro, Galicia, northwestern Spain.

Figure 10.12. (a) Clastic fragment of Opal-A. (b) Opal-A coats the porous texture near the tip of speleothem from Traba Mountains (northern Galicia, Spain). (c) Longitudinal section of a speleothem from a sandstone shelter on the Tindal Plain (Northern Territory).

Figure 10.13. Anomoeoneis (diatom) from Tcharkuldu Hill, Eyre Peninsula, South Australia buried by opal-A precipitation.

massive opal gel precipitation (Fig. 10.13). This gel acts, during the third phase, as porous substratum allowing the growth of low energy crystallisation whiskers of pure minerals as phosphates, calcium carbonate, but habitually gypsum with an origin in bacterial activity. The morphology of these crystals is determined by the space characteristics to grow (normally the drop size and shape) where they develop. The crystals show different habits: crest and rosettes, needles, prismes, etc. (Fig. 10.14). They are most frequently developed in the end of the speleothem

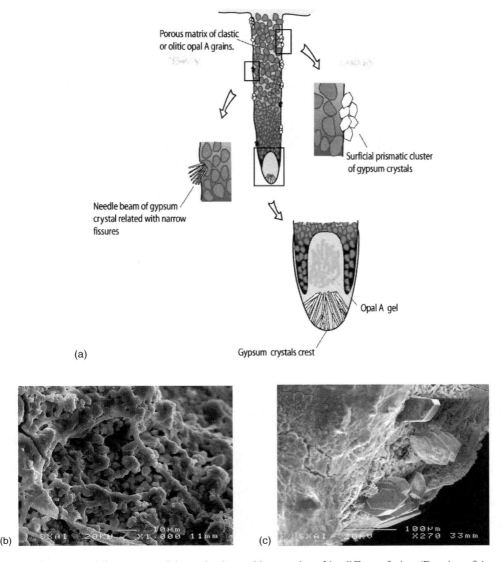

Figure 10.14. (a) Different parts of the speleothem with examples of its different facies. (Drawing of A. Grandal d'Anglade). (b) Detail of porous matrix of clastic or oolitic opal-A grains. (c) Surficial prismatic cluster of gypsum crystals.

appearing as a radial association of crystals with planar development giving rise to a typical cauliflower morphology. The first known mention of speleothems in granites corresponds to Caldcleugh (1829) who reported silica speleothems from an overhang in a granite-gneiss dome in the Rio de Janeiro area, Brazil, but speleothems are reasonably commonplace having been reported from granite outcrops in Spain and Portugal, in the Black Forest of southern Germany, in Brittany, in Madagascar, in U.S.A. and Córdoba and Anillaco (República Argentina); in quartzites and sandstones having been reported in the Hoggar Mountains of the central Sahara, in Roraima Plateau of Venezuela, in the Sydney Basin, New South Wales and in Litchfield National Park, near Darwin, Northern Territory.

Figure 10.14. (d) Needle-beam of gypsum crystals related with narrow fissures. (e) Opal-A infilled fabric, and (f) Gypsum crystal crest.

REFERENCES

Bennett, P.C. 1991. Quartz dissolution in organic-rich aqueous systems. *Geochimica et Cosmochimica Acta* 55: 1781–1797.
Boyé, M. and Fritsch, P. 1973. Dégagement artificiel d'un dome crystallin au Sud-Cameroun. *Travaux et Documents de Géographie Tropicale* 8: 69–94.
Bradley, W.C., Hutton, J.T. and Twidale, C.R. 1978. Role of salts in development of granitic tafoni, South Australia. *Journal of Geology* 86: 647–654.
Caldcleugh, A. 1829. On the geology of Rio de Janeiro. *Transactions of the Geological Society* 2: 69–72.
De Prado, C. 1864 (1975). Descripción física y geológica de la Provincia de Madrid. Publicaciones especiales Colegio de Ingenieros de Caminos Canales y Puertos, Madrid, 352 p.
Dorn, R.I. and Oberlander, T. 1981. Microbial origin of desert varnish. *Science* 213: 1245–1247.
Drewry, D. 1986. Glacial Geologic Process. Edward Arnold, London.
Engel, C.G. and Sharp, R.P. 1958. Chemical data on desert varnish. *Geological Society of America Bulletin* 69: 487–518.
Finlayson, B. and Webb, J. 1985. Amorphous speleothems. *Cave Science* 12: 3–8.
Hult, R. (1873). Fran Nord till Syd Kalender Fjfillvandringar i Galicien och Zamora. *Geografiska Foreningen i Finland* 30–55.
Ikeda, H. 1990. Tafoni topography and its development process as seen in the Jhumonjin area of the northeastern coastal Korean Peninsula. *Memoirs of Nara University* 18: 49–66.
Ikeda, H. 1994. Topography study of a granite cave – the example of a cave discovered in a mountain in the northern part of Seoul, Korea. *Memoirs of Nara University* 22: 1–14.
Jutson, J.T. 1917. Erosion and the resulting landforms in sub-arid Western Australia. *Geographical Journal* 50: 418–437.
Kastning, E.H. 1976. Granitic karst and pseudokarst, Llano County, Texas, with special reference to Enchanted Rock cave. Proceedings 76th National Speleological Society, Annual Convention, 43–45.
Klaer, W. 1956. Verwitterungsformen im Granit auf Korsika U.E.B. Haack Gotha.

Meybeck, M. 1987. Global chemical weathering of surficial rocks estimated from river dissolved loads. American *Journal of Science* 287: 401–428.

Mustoe, G.E. 1982. The origin of honeycomb weathering. *Geological Society of America* 93: 108–115.

Penck, A. 1894. Morphologie der Erdoberfläche. Engelhorns, Stuttgart. 2 volumes.

Reusch, H.H. 1883. Notes sur la geologie de la Corse. *Société Geologique de France Bulletin* 11: 53–67.

Scheffer, F., Meyer, B. and Kalk, E. 1963. Biologische Ursachen der Wüstenlackbildung. *Zeitschrift für Geomorphologie* 7: 112–119.

Thomson, J. 1863. On the disintegration of stones exposed in buildings and otherwise exposed to atmospheric influences. British Association for the Advancement of Science, 32nd Meeting, Notes and Abstracts Murray, London.

Vidal Romaní, J.R. and Vilaplana, J.M. 1984. Datos preliminares para el estudio espeleotemas en cavidades graníticas. *Cuadernos do Laboratorio Xeoloxico de Laxe* 7: 305–324.

Vidal Romaní, J.R. 1983. El Cuaternario de la provincia de La Coruña. Geomorfología granítica. Modelos elásticos de formación de cavidades. Tesis Doctoral. Publicaciones Universidad Complutense de Madrid.

Vidal Romaní, J.R. 1998. Las aportaciones de Casiano de Prado a la geomorfología granítica. *Geogaceta* 23: 157–159.

Vidal Romaní, J.R. and Twidale, C.R. 1998. Formas y paisajes graníticos. Servicio de Publicaciones de la Universidad de Coruña. Serie Monografias, 55. Coruña, España.

Vidal Romaní, J.R., Twidale, C.R. and Bourne, J. A. 1998. Espeleotemas y formas constructivas en granitoides. In Investigaciones Recientes de la Geomorfología Española. Actas V Reunión Nacional de Geomorfología. Granada, España. pp. 772–782.

Vidal Romaní, J.R., Twidale, C.R. and Bourne, J. A. 2003. Siliceous cylindrical speleothems in granitoids in warm semiarid and humid climates. *Zeitschrift für Geomorphologie* 47(4): 417–437.

Wilhelmy, H. 1964. Cavernous rock surfaces (tafoni) in semiarid and arid climates. *Pakistan Geographical Review* 19: 9–13.

11

Split and cracked blocks and slabs

Many granite blocks, boulders, slabs and plates have been split as a result of the development of fractures. Some are isolated, others arranged in distinct, if in some instances irregular and varied, patterns. Some are open, others tight. Some are joints, others faults. Some of the faults have been active recently, and minor fault scarps, occurring either singly or in regular arrangements to give minor horsts and grabens, are commonplace on granite outcrops (Fig. 11.1). In other instances, however, the displacements are much greater.

11.1 SPLIT ROCKS

11.1.1 *Description*

Some boulders have been split in two parts (Fig. 11.2). In many instances the two parts are of equal mass, or nearly so. The parting fractures are most commonly planar, but some are arcuate or wavy.

Figure 11.1. (a) Minor fault scarps Minnipa Hill, northwestern Eyre Peninsula, South Australia, formed 19 January 1999.

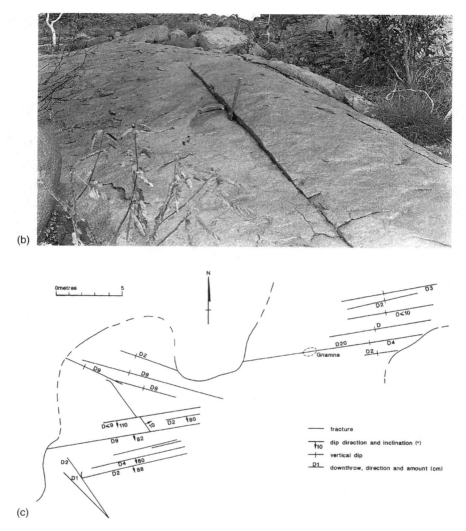

Figure 11.1. (b) Minor scarp on low dome, western Pilbara, Western Australia. (c) Horst and graben delineated by minor fault scarps, The Granites, Mt Magnet, Western Australia.

Some of the partings are merely latent or secondary fractures for they conform to the regional fracture pattern (Fig. 11.3), but others cut across pre-existing structures. Split rocks occur in arid lands such as the Australian deserts, as well as such humid tropical regions as the Tampin area of West Malaysia (Tschang, 1962), in temperate regions like the Mt. Lofty Ranges and Eyre Peninsula, South Australia, as well as nival areas like the Pyrenees, the uplands of the Iberian Peninsula and the Kosciusko region of southeastern New South Wales (Figs 11.2 and 11.3). In all areas the splitting affects exposed rocks.

11.1.2 *Origin*

Several writers have attributed split rocks to heating and cooling under hot desert conditions. Whitaker (1974) records the splitting of rock near Halls Creek, in northwestern Australia, in 1952, shortly after a rainstorm. Unfortunately, it is not known whether the split developed along a

Spilt and cracked blocks and slabs 261

Figure 11.2. Split boulders at (a) Devil's Marbles, Northern Territory; (b) in the Mt Kosciusko area of New South Wales, showing a split divergent from an earlier injected vein or sill.

pre-existing latent joint. In any case, the occurrence of split rocks in cold and warm humid regions precludes this mechanism as of general application. Also, bearing in mind the large volumes of rock involved, the essentially superficial nature of insolation changes, and the poor conductivity of rocks, it seems doubtful whether heating and cooling alone, even aided by rain showers, could achieve the splitting of large homogeneous masses, though cobbles and pebbles are evidently so affected. Some writers attribute split rocks to pressure release, but the geometry of the split blocks, their size and likely accumulated stress, and the general history of boulder development render this suggestion unlikely.

262   *Landforms and Geology of Granite Terrains*

Figure 11.2.   (c) In the Tampin area (x) West Malaysia.

Figure 11.3.   Split rocks exploiting latent partings conformable with the regional fracture patterns (a) in the Pyrenees.

The most plausible general explanation of split rocks is based on gravity. Secondary fractures (i.e. shears not involved in delimiting the joint blocks) within the original mass (Fig. 11.4) or rift and grain (Chapter 2) are exploited by moisture attack either in the subsurface, concurrently with the differential weathering that produces corestones (see Chapter 5) or after exposure of the corestones as boulders. On exposure, the corestones are no longer confined and supported, and though the boulders and blocks are defined and delineated by orthogonal joints, it is a matter of observation that they include other secondary or latent joints. Even if the exposure of the corestone is incomplete, the spheroidal mass is subjected to tensional stress rather than compressional and gravitational (lithostatic) loading.

Figure 11.3. (b) Near Chazeirollettes, southern Massif Central, France.

Figure 11.4. Boulder with secondary fracture, unexploited by weathering or gravity, Hyden Rock, southern Yilgarn, Western Australia.

Subsidiary fractures are penetrated by moisture and so widened, either as a result of freeze-thaw activity or ice-wedging (as in the Pyrenees and the Snowy Mountains of New South Wales, for example), or in consequence of alteration of the rock immediately adjacent to the parting. Weathering along the secondary fracture weakens cohesion between the two adjacent masses, and, unless the blocks and boulders have flat bases and rest on an even platform, the weight of the two parts of the mass separated by the latent fracture causes them to fall apart (Fig. 11.5). Seismic shaking may have assisted the process.

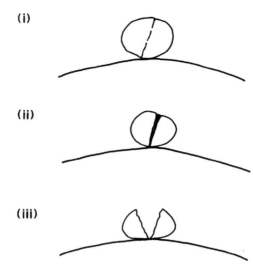

Figure 11.5. Suggested development of split rock.

The process may be initiated in the subsurface though splitting cannot occur until after exposure. Given a corestone set in grus, evacuation of the matrix exposes the corestone as a boulder. Instead of being supported on all sides by grus the weight of the boulder rests only on the still buried part, and the stress regime changes from one of compression to one of flexure, with tension on the upper side of the boulder and compression on the lower (Fig. 11.5). Any latent steeply-inclined fractures, or other planes of weakness, thus tend to be pulled apart.

Some blocks and boulders have been split twice or more (Fig. 11.3a). For these, frost riving affecting secondary, latent fractures offers the most satisfactory explanation, for all occur in areas subject to frequent freeze-thaw alternations, either at present or in the recent past. Even so, many sit on a foundation too even and stable for the separate blocks to fall apart.

11.2 PARTED AND DISLODGED BLOCKS

Some blocks defined by steeply inclined fractures have apparently been joggled or moved (Fig. 11.6), in places, as at Kokatha, in the northernmost Gawler Ranges, South Australia, by as much as 2 m and on Dartmoor, southwestern England, by more than a metre. Such parted blocks originally consisted of two cubic or quadrangular blocks that were, presumably, closely juxtaposed, but which are now separated by a considerable gap. In some instances, both gap and marginal blocks are surmounted by other blocks.

These forms cannot be accounted for by unbuttressing and the two adjacent blocks rotating outward in opposed directions, as suggested in explanation of split boulders. The blocks in question are angular and could not roll apart, even if the blocks to either side were eliminated. In some instances, the parted blocks are surmounted by other large residuals, the weight of which would surely prevent the underlying masses from moving. Worth (1953) mentions lightning in relation to such dislodged blocks, though the mechanism is not detailed, and there is in any case no evidence suggestive of such strikes.

The parted blocks are not due to the weathering and erosion of sills of weaker rock, for no sign of such materials is found in association with them. They cannot be attributed to the slippage of blocks under gravity, for the parted blocks characteristically stand on sensibly horizontal bases. Swelling clays that are capable of moving considerable loads are rarely, and certainly not necessarily, present at the appropriate sites, and massive frost wedging cannot be cited as an explanation

Figure 11.6. Dislodged blocks involving lateral dislocation at (a) Christopherus-Stein, Natural Park Blockheide Gmünd-Eibenstein, Gmünd, Lower Austria (human scale 1.72 m high) (courtesy of K.H. Huber). (b) On Dartmoor, southwestern England.

Figure 11.6. (c) On Dartmoor, southwestern England.

of all parted blocks, for, though it could conceivably have operated on Dartmoor, southwestern England, and in the Serra do Gêrez, Galicia, NW Spain, for instance, during the Pleistocene, it can safely be ruled out for the low altitude tropical examples of parted blocks.

On the other hand, and as it has been mentioned, many granite domes and platforms, even those in the comparatively stable cratonic areas, have been joggled, for small fault scarps, some arranged in horst and graben patterns (Fig. 11.1), are commonplace, and A-tents and areas of chaotically disposed slabs occur in association with some of the displaced blocks (see below). Thus, while some dislodged blocks may be due to the action of ground ice, others can be attributed to seismic shaking. Most, however, remain enigmatic.

11.3 DISLOCATED SLABS

Some granite forms consist of slabs and blocks that have been split, cracked and displaced. They are all developed in intrinsically fresh bedrock. By far the most common member of this suite is the A-tent. Others are overlapping slabs, displaced slabs, and vertical and horizontal wedges. The blisters named by Blank (1951) appear to be related to A-tents, but their crests are not fractured.

11.3.1 *A-tents*

A-tents consist of pairs of slabs, frequently roughly rectangular in shape, each touching the adjacent rock surface at their outer extremities, but standing a few centimetres above the general surface level where they are in contact in a fracture zone known as the roof line (Fig. 11.7a). They delineate triangular cavities, as do ridge tents: hence their name. In some areas one or both of the constituent slabs are split by a fracture disposed obliquely to the main crestal fracture.

A-tents, pop-ups in North America (Rutty and Cruden, 1993; Wallach, Arsalan, McFall *et al.*, 1993) and chapitaux by French workers, are well-known in Australia (Twidale, Campbell and Vidal Romaní, 1993; Twidale and Bourne, 2000), being abundantly developed, for example, on inselbergs on Eyre Peninsula, South Australia, the Kulgera Hills, Northern Territory, and the Yilgarn and Pilbara blocks, Western Australia (Plug and Plug, 1997 and see Fig. 11.7b). They occur in many parts of northern Africa (Boissonnas, 1974). They are reported from Labrador, central Texas and Georgia, from the Rupununi savannas of Guyana and the Sabah district of east Malaysia.

Spilt and cracked blocks and slabs 267

Figure 11.7. (a) Section of A-tent. (b) A-tent on Kokerbin Hill, Western Australia.

A-tents typically occur on midslope (Fig. 11.8a). They are up to 6 m long and up to almost 8 m wide, but most are smaller, means of 2 m long and 1 m wide being more typical. All A-tents have a crestal fracture and most are also fractured at their lateral terminations; but in some the slab extends with no apparent break into the adjacent hillslope. Most of the A-tents that have been studied are 10–15 cm high and involve slabs about 10 cm thick, but one on Wudinna Hill, northwestern Eyre Peninsula, South Australia, has slabs 580 mm thick standing 820 mm above the floor (Fig. 11.8b; also X in Fig.11.9), and one on Carappee Hill, on northern Eyre Peninsula, comprises slabs only 13 mm thick (Fig. 11.8c). A beam of granite some 15 m long, just over 2 m wide at its broadest point, 79 cm thick maximum, standing up to 15 cm higher than the adjacent slope (Fig. 11.10a) and located on Wudinna Hill, was formerly construed as a vertical wedge, caused by squeezing between adjacent slabs in a direction normal to the length of the beam. It was first interpreted (Jennings and Twidale, 1971) as due to NE-SW compression and direct contact squeezing by the slabs on either side. But the sides of the wedge are not (now?) in contact with the adjacent walls, and the feature is more plausibly interpreted as an elongate A-tent due to compression from NW and SE, in line with other A-tents on the hill and in the vicinity. Many small triangular wedges occur along fractures on a disturbed sector of the adjacent slope (W in Fig. 11.9). In most areas the crests of A-tents are aligned (Figs 11.11a and b), but in others their orientation varies, in some instances being aligned normal to each other at adjacent sites (Fig. 11.11c). The terminal and crestal fractures typical of the angular forms are not present in blisters, formed by arched shells or slabs, typically 1–2 cm thick. There is a continuum of forms between thin sheets that sound hollow when tapped, through arched slabs which lack crestal cracks, arches with incomplete or discontinuous cracks, and A-tents. Most angular A-tents occur on midslope sites, whereas arches are

268 *Landforms and Geology of Granite Terrains*

(a)

(b)

Figure 11.8. Large A-tents (X) on western midslope of Wudinna Hill, northwestern Eyre Peninsula, South Australia: (a) general view; (b) detail.

Spilt and cracked blocks and slabs 269

Figure 11.8. (c) Thin A-tent in granite gneiss on western midslope of Carappee Hill, northeastern Eyre Peninsula, South Australia.

Figure 11.9. Plan of part of western flank of Wudinna Hill, showing large A-tent (X) shown in Fig. 11.8b, new A-tent (Y), vertical wedges (W) and area of chaos (Z).

270 *Landforms and Geology of Granite Terrains*

Figure 11.10. Vertical wedge, or elongate A-tent, located low on western slope of Wudinna Hill, Eyre Peninsula, South Australia.

found both there and on the crests of hills. Arched slabs are fairly commonplace in tropical regions. At Cash Hill, northwestern Eyre Peninsula, South Australia, an A-tent was, in the early seventies, partly covered by regolith capping the upland. It is possible that the A-tent had recently been exposed by the erosion of the regolith, and that it had developed beneath the land surface. The feature now stands adjacent to the regolithic cover (Fig. 11.12).

Measurements of A-tents from several sites on northern Eyre Peninsula suggest that, if, as seems certain, the presently raised slabs that constitute the A-tents were originally part of the smooth hillside; their present combined lengths would in most instances exceed the space they originally filled by some 3–4%. One on Lightburn Rocks in the Eastern Great Victoria Desert of South Australia suggests an expansion of 5%. On the other hand, some measured at The Granites, near Mt Magnet, in Western Australia, imply an expansion of only 1%.

Most A-tents predate human occupation or at least the recording of events, but some have formed recently. One formed on the lower western midslope of Wudinna Hill between February and May 1985 (Y in Fig. 11.9 and Fig.11.13a), and two have developed at the nearby Quarry Hill, one, a complex series of buckles, following a detonation early in 1993, the other on a stripped surface early in 1995 (Figs 11.13b and c).

Spilt and cracked blocks and slabs 271

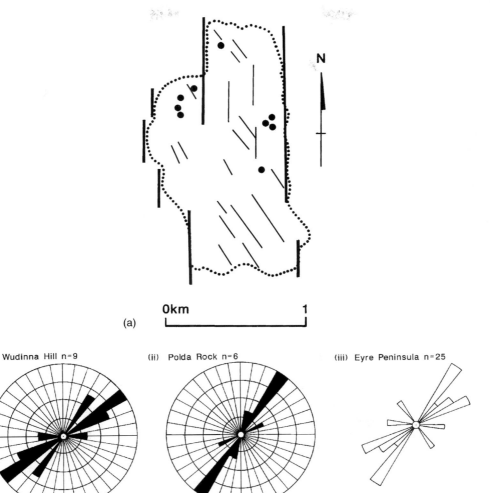

Figure 11.11. (a) Plan of Wudinna Hill, Eyre Peninsula, South Australia, showing major fractures (heavy-postulated, light-mapped) and location of A-tents and related forms. (b) Plots showing orientation of crests of A-tents at two sites on northwestern Eyre Peninsula, and for all of the area.

11.3.2 *Overlapping slabs*

On the northwestern midslope of Wudinna Hill there is an overlapping slab consisting of a plate of granite, the upper end of which is raised and overlaps another that forms the adjacent upslope section of the hillside by about 300 mm (Fig. 11.14). The two irregular edges of the raised and the flat slabs are disposed roughly normal to the surface and match perfectly, suggesting that they were once juxtaposed.

11.3.3 *Displaced slabs*

Perhaps the most spectacular example of a displaced slab is developed on Little Wudinna Hill, near Wudinna, northwestern Eyre Peninsula, South Australia (Twidale, Schubert and Campbell, 1991). Here, a triangular sheet some 410 mm thick and 9 m long has slipped some 8.5 m down a 16° slope, coming to rest against the steepened and flared lower slope of the hill (Fig. 11.15a).

272 *Landforms and Geology of Granite Terrains*

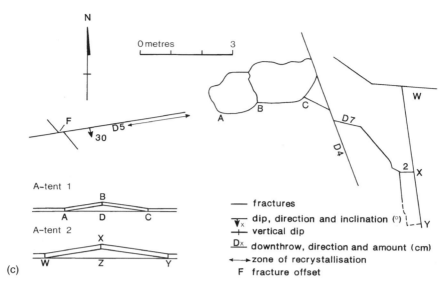

Figure 11.11. (c) Plan of A-tents with crests disposed normal to one another at The Granites, Mt Magnet, Western Australia. Note fault dislocation.

Figure 11.12. Small A-tent on Cash Hill, northwestern Eyre Peninsula, located close to edge of regolith and possibly initiated beneath formerly more extensive spread of detritus.

The triangular gap whence came the slab remains identifiable. Several of the slabs bordering the gap display a zig-zag pattern in plan due to the development of brittle tensional fractures (Fig. 11.15b). Several more, similar, slabs remain in a tumbled mass below the triangular one. Above the gap left by the slabs, several of the sheets of rock have slipped a few centimetres downslope. However, the complex A-tent (X) developed some 20 m upslope from the major slippage is not associated with any downslope movement (Fig. 11.16). The uplift of the A-tent slabs has, however, caused some

Figure 11.13. Recently formed A-tents: (a) on western midslope of Wudinna Hill, northwestern Eyre Peninsula, South Australia, and formed between February and May 1985; (b) on Quarry Hill, near Wudinna, northwestern Eyre Peninsula, South Australia: one formed in early 1993.

adjacent slabs to be disturbed; in particular some have been pushed sideways, one of them now impinging by some 25 cm into the gutter that delineates the slope on its southern side (Y in Fig. 11.16). On nearby Wudinna Hill, a slab roughly 5.4 m by 7.5 m and 320–530 mm thick has moved downslope 230–260 mm on a 19° incline, and a similar slippage is evidenced on one of the Kulgera Hills, Northern Territory (Figs 11.17a and b). Displaced blocks have been noted in many other places (e.g. Fig. 11.17c).

Displaced slabs are anomalous forms because there are many examples of blocks and boulders perched precariously on slopes steeper than those on which slippage of slabs has occurred. The

274   *Landforms and Geology of Granite Terrains*

Figure 11.13.   (c) One formed in early 1995 on a bedrock surface from which about a metre of regolith had recently been stripped.

Figure 11.14.   Overlapping slab, Wudinna Hill, northwestern Eyre Peninsula, South Australia.

inclination and roughness of the surfaces in contact is crucial, but in a material like granite is of the order of 30°. Thus, at Pildappa Rock, northwestern Eyre Peninsula, there are large, sub-angular joint blocks, more than one metre diameter, standing on slopes of 24° and 26° (Fig. 11.18). Though the slopes on which they rest are steeper than those down which the triangular slab at

Spilt and cracked blocks and slabs   275

Figure 11.15. (a) Slipped triangular slab, Little Wudinna Hill, northwestern Eyre Peninsula, South Australia, and (b) tension fractures at edge of space vacated by displaced slab.

Little Wudinna Hill has slipped, and though the granite is of the same order of roughness, the blocks standing on them are stable. The reason is that, though in general view smooth, the granite slopes and the surface of the granite slabs and blocks themselves are, in detail, very rough or pitted. A micro-relief of a few millimetres has developed and, even when lubricated by water, friction inhibits slippage on these slopes. Yet some slabs have clearly migrated under gravity on slopes that are more gentle than those on which blocks stand firm.

11.3.4 *Chaos*

Displaced slabs in chaotic arrangement have been noted at several sites associated with A-tents, as for instance on the western side of Wudinna Hill (Z in Fig. 11.9), on Quarry Hill, near Wudinna,

276 *Landforms and Geology of Granite Terrains*

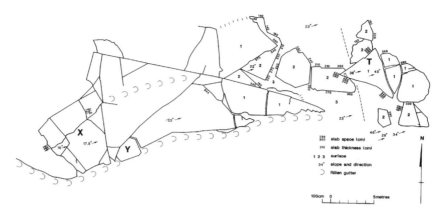

Figure 11.16. Plan of slope of Little Wudinna Hill (Fig. 11.15a), showing slipped triangular slab (T), A-tent (X) and laterally displaced slab (Y).

Figure 11.17. Slipped slabs (a) Wudinna Hill, northwestern Eyre Peninsula, South Australia.

Eyre Peninsula, South Australia (Fig. 11.19), and on a low hill just north of Payne's Find, Western Australia. In part, such chaos may be due to the collapse of A-tents, but this does not afford a complete explanation, and recourse needs be made to shaking, disruption and displacement of pre-existing shells of rock.

11.3.5 *Wedges*

Wedges of triangular cross-section are fairly commonly developed at the exposed lower edges of sheet structure. Wedges are not, of course, slabs, but it is convenient to discuss them here together with other members of this suite of dislocated forms.

Several wedges have been displaced laterally (Fig. 2.9d) though some remain *in situ* (Fig. 11.20). Dislocation and friction have caused wedges of triangular cross section, and typically a few metres long, to break away from the lower exposed faces of sheeting slabs. Some remain *in situ*, but many

Spilt and cracked blocks and slabs 277

Figure 11.17. (b) Kulgera Hills, Northern Territory, Australia. (c) Slipped block, Kokerbin Hill, southern Yilgarn of Western Australia. (d) Slab on Minnipa Hill, northwestern Eyre Peninsula, South Australia, displaced on 19 January 1999.

Figure 11.18. Perched slabs and blocks on upper slopes of Pildappa Rock, northwestern Eyre Peninsula.

(a)

Figure 11.19. (a) Chaos of thin slabs on slope of Quarry Hill, near Wudinna, northwestern Eyre Peninsula, South Australia.

have either been dragged or have fallen a few centimetres or metres downslope. Vertical wedges are due to compression and the rupture and vertical displacement of slivers of rock, triangular in cross section, marginal to fractures, typically fractures of the orthogonal system. Several recently formed examples have been plotted on the flanks of Wudinna Hill, Eyre Peninsula (W in Fig. 11.9a).

11.3.6 Origin of the forms

Some displaced slabs and A-tents have been attributed to insolational heating and to offloading consequent on erosion. They could also be explained in terms of rupture following downslope slippage and impact against an immovable block or surface, or disruption by tree roots.

Figure 11.19. (b) Chaos of thin slabs on Minnipa Hill, near Wudinna, northwestern Eyre Peninsula, South Australia, formed on 19 January 1999.

Figure 11.20. Horizontal wedge *in situ*, Ucontitchie Hill, northwestern Eyre Peninsula, South Australia.

Angular buckles have been attributed to insolation. But the Sun's heat can be discounted as the cause of the A-tents and the other forms considered here, because insolation argues heating and cooling, and hence a reversible event, whereas the A-tent represents an irreversible change in the volume of the granite involved. Displacement by such heating and cooling is conceivable, but permanent deformation is not. If insolation were responsible, A-tents ought to occur preferentially on sunny aspects, such as crests and western slopes, but no such distribution has been noted. The ephemeral though intense heat generated in forest, scrub or bushfires undoubtedly causes flaking and spalling of exposed surfaces, but the bedrock surfaces on which A-tents are formed lack vegetation so that there is no fuel for fires. Also, the flakes produced by such heating are typically thinner and less regular than the slabs involved in the thicker A-tents. Furthermore, there is some suggestion that A-tents are initiated in the subsurface beyond the range of either insolation or the heat of bushfires.

If A-tents were due to offloading (see Chapter 2), their crests would be aligned parallel to the contour of the slope on which they occur, whereas in many areas they display a preferred orientation which is geometrically related to regional tectonic trends (Fig. 11.11). The suggestion that A-tents are a manifestation of pressure release consequent on erosional offloading does not explain why the angular A-tents are typically developed on midslope, and it is not consistent with the survival of the host masses of granite in inselbergs which have resisted weathering and erosion, because they are massive, monolithic and in compression. Two of the three A-tents of recent origin noted in the Wudinna district have formed without the aid of erosion, either natural or artificial, while the third is on a surface exposed by the removal of about a metre of grus.

A-tents and associated forms are manifestly youthful, yet occur on host forms that are evidently of some antiquity; which is inconsistent, for if due to pressure release the minor suite would surely have formed as soon as the host forms were delineated by differential subsurface weathering.

The suggestion that tree roots have, during their growth and thickening, raised slabs of rock to such an extent that friction has been overcome, is evidenced at many sites. Vertical displacement of granite slabs has taken place at one site on Wudinna Hill, but no lateral displacement is involved, the displacements are irregular, and presumably the slabs will return to, or close to, their original positions once the tree dies and decays.

If A-tents were due to impact following slippage, there ought to be gaps whence came the slabs upslope from the A-tents. There are no such spaces associated with A-tents (e.g. Fig. 11.9). Conversely, where slippages do occur they are not associated with A-tents. Slipped slabs and A-tents are associated and may have a similar causation, but they are not genetically related.

Some mineral alteration causes expansion, and it could be argued that such expansion could account for A-tents. Certainly, some of the granite in which A-tents are well-developed, for instance at Mt Magnet, Western Australia, Kulgera, Northern Territory and Augrabies, Namibia, is impregnated with iron, manganese and silica, but many other A-tents are developed in intrinsically fresh granite. Moreover, the solution of limestone (in which A-tents are developed – see Chapter 12) seems unlikely to result in volume increase.

Overlapping slabs cannot be explained as due to slippage, for the lower of the two components involved laps over the upper and not the reverse, as would be the case if downslope movement were involved. Furthermore, there are no gaps at either the lower or upper end of the two slabs involved. The feature could be a collapsed A-tent, but the collapse has not been caused by attrition of the rock near the erstwhile crestal fracture, because the two opposed ends of the slab match perfectly, and the rock shows no signs of undue weathering. If an A-tent collapsed with the downslope slab coming to rest on the upslope member, it must have been caused by some catastrophic event. A similar overlapping slab occurs on the northeastern midslope of Little Wudinna Hill. It now overlaps the adjacent slab a matter of 15 cm, and another platy fragment has slipped beneath the raised slab. That A-tents are associated with shaking is suggested by an A-tent located high on the eastern slope of Wudinna Hill. Here the original arrangement of slabs could not be reconstituted even if the effects of expansion were removed, for a third slab, formerly located just upslope, has slipped into the cavity formed by the raised slabs of the A-tent (Fig. 11.21).

On the other hand, many parts of the world are known to be in substantial compression. Earthquakes and earth tremors are ubiquitous. Dramatic disturbances are associated with some. Thus, large boulders, originally partly embedded in regolith, but in contact with other boulders beneath the surface, were squeezed up (Fig. 11.22) and translocated laterally some 40 cm during the Great Hanshin earthquake of January, 1995 (Ikeda, 1996). The horizontal acceleration of pressure waves associated with major earthquakes is known to buckle artificial paving, and shock waves generated by quarry blasting are known to cause A-tents and other disturbances (Twidale and Sved, 1978). Similar contemporary arching associated with rock bursts is reported in a granite quarry in Tocumweal, New South Wales. Again, in the Mariz Quarry near Guitiriz, Galicia, both vertical and horizontal wedges, as well as small large-radius domes, can be seen developed in association with compressive structures (see Chapter 2 and Fig. 2.9).

Figure 11.21. A-tent with displaced slab beneath it, high on eastern slope of Wudinna Hill, northwestern Eyre Peninsula, South Australia.

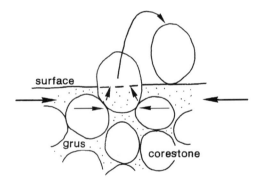

Figure 11.22. Suggested mechanism for translocation of boulder during the Kobe-Mt Rokka earthquake of January, 1995.

At Mt Magnet, one A-tent is demonstrably associated with fault dislocation (Fig. 11.12) and one large, and many small, A-tents formed during an earth tremor (2.3 on the Richter Scale), which affected part of Minnipa Hill, on northwestern Eyre Peninsula, on 19 January 1999. Another line of evidence favouring compressive stress, and difficult to refute, concerns developments noted during the quarrying of Palaeozoic limestone in Ontario. Coates (1964) has reported that the floor cracked and heaved, forming an A-tent the crest of which stood 2.4 m higher than it did originally, and affecting rocks some 15 m to either side of the crestal fracture. Calculations showed that pressure release could not quantitatively account for the upheaval, which was attributed to horizontal stress in the limestone. The forms described in this section are readily explained as being associated with the release of compressive stress, particularly if any remnant stress due to earth movements were ephemerally, but critically, increased during earth tremors. The horizontal components of earthquake motion play an important part in the initiation of landslides and, in general terms, the sudden release of strain energy in the rock resulting from a high horizontal stress field could induce both failure of the surface rock and sufficient shaking to produce instability.

Such an explanation accounts for the varied results of A-tent development involving expansion, and for the preferred location of angular A-tents and other forms described here on midslope where stresses are greatest on an arched (convex) surface. Variations in crest orientation can be explained in terms either of shear couples or of separate periods of stress from different directions. Arched and angular forms are in these terms members of a continuum. Given that the near-surface granite is subdivided into layers (it is divided by sheeting joints and flaggy partings), lateral pressure may result in arching. Further compression would cause a crestal fracture to develop, and still greater pressure, the marginal fractures. Alternatively, A-tents may evolve during a single compressive event.

Calculations of the internal friction of slabs involved suggest that A-tents are precariously stable under the influences of gravity and pore pressure, and that increases in horizontal stress such as are generated in earthquakes and tremors could cause instability and buckling.

If the forms are related to a catastrophic event or events such as earthquakes or tremors, triangular wedges can be explained in terms of a sudden, possibly short-term, increase in compressive stress resulting in differential movement along sheeting planes, and frictional drag, fracture and dislocation on the exposed sheeting plane, very much as slippage along bedding planes during folding can produce slickensides, recrystallisation, etc. The resultant shaking would explain how both friction and inertia have been overcome, as is implicit in slipped slabs. Increased horizontal compression would also account for the buckling and expansion implicit in the formation of A-tents. It is notable that on Lightburn Rocks where A-tents are well-developed, there are also many recently developed fractures, including some vertical dislocations between juxtaposed slabs. Also, the assemblage of forms discussed here is well-developed on northern Eyre Peninsula, in a zone bounded by faults and known to be seismically active. The late Dr D.J. Sutton reported that a small portable seismograph set up on Mt Wudinna for twelve hours in late 1975 recorded two seismic events in that short period. The displaced slabs on Quarry Hill are due to blasting – an artificial miniature earthquake.

11.3.7 *Relationship of A-tents and pressure ridges*

The A-tents in sandstone developed in Wyoming, and illustrated in Scott (1897), form an elongate ridge, and the question arises as to the relationship between A-tents and pressure ridges developed along fault traces such as those of the San Andreas Fault in California (Thomas, Wallach, McMillan *et al.*, 1993). In particular, are A-tents and the linear ridges described from southern Ontario and from the bed of Lake Ontario comparable? In broad terms, both appear to be due to tectonic pressures, but first, judging by illustrations of those formed on land in eastern Canada, the ridges developed there are not hollow but consist of strata pushed up against each other and upthrust, rather like miniature horst structures; and they develop along considerable lengths of the fault trace, presumably wherever compressive stress is enough to rupture the strata. Second, the linear ridges are due to direct pressure of one block against another, as are the vertical wedges plotted on the flank of Wudinna Hill and noted in Fig. 11.9. In A-tents, on the other hand, the pressures responsible for the buckling are transmitted and affect only pre-existing suitable situations, namely a lamination of the near-surface rock. The difference can be illustrated in relation to the rock beam or elongate A-tent illustrated in Fig. 11.10. It can be interpreted as either having been squeezed up by direct pressure from the adjacent blocks (as have the wedges plotted on Fig. 11.9); or, and this is the construction favoured here, raised because of pressure normal to the beam and imparted at either end.

11.4 POLYGONAL CRACKING

11.4.1 *Description*

Isolated and randomly disposed groups of cracks occur on many granite surfaces. They have been called hieroglyphs (Fig. 11.23a). Some, located on the convex swells of boulders, describe radial patterns (Fig. 11.23b). Elsewhere, and more commonly, however, the cracks occur in distinct and repeated arrangements. Polygonal cracking or weathering consists of a pattern of cracks, some as much as 5 cm wide developed in a superficial shell or shells of rock: they extend to no more than a few centimetres beneath the surface of boulders, blocks or platforms. They form orthogonal,

(a)

Figure 11.23. (a) Hieroglyphs on granite surface at Daadenning Hill, near Merredin, southwest of Western Australia.

(b)

Figure 11.23. (b) Radial fractures on boulder at Corrobinnie Hill, northern Eyre Peninsula, South Australia.

284 *Landforms and Geology of Granite Terrains*

Figure 11.24. Polygonal cracking: (a) orthogonal, at Buccleugh, near Johannesburg, South Africa; (b) on boulder at Tcharkuldu Hill, northwestern Eyre Peninsula, South Australia.

rhomboidal, or polygonal patterns (though some are irregular or crazy, as in crazy paving) defining thin plates (Fig. 11.24). Various stages in the development of the cracks have been recognised. The juvenile narrow fractures are linear, and the edges cut through individual crystals, but those widened by weathering are in detail jagged due to the protection of individual crystals, particularly of quartz. The polygonal plates range in diameter from 2 cm to some 24 cm, with the average and

Figure 11.24. (c) On rock platform, Corrobinnie Hill, northern Eyre Peninsula, South Australia.

mode both near the upper end of the range. Some of them are slightly curved or convex upward in respect to the surface of the host rock. On boulders on Tcharkuldu Hill, on Eyre Peninsula, South Australia and at The Granites, near Mt Magnet, Western Australia, some of the thin plates have been either worn away or have fallen, so that as many as three layers, each with polygonal cracking developed, are exposed. The narrow hairline cracks of the deeper layer stand in marked contrast with the wider, less sharply defined, fractures of the outer skin.

The granitic rocks in which the most clearly defined and prolific polygonal cracking occurs are characteristically equigranular and medium grained, though examples on fine grained aplitic rocks have also been noted. Thus, the granite at Tcharkuldu Hill, where the cracking is most abundantly developed, is of this type. Conversely, where the granite is porphyritic, as, for instance, at Wudinna Hill, no polygonal cracking has been observed. Such features are formed at the weathering front. Commonly, though not universally, polygonal cracking is associated with heavy surficial indurations of iron oxide, manganese oxide, and possibly silica. This association is obvious at such notable granitic polygonal cracking sites as Tcharkuldu Hill, The Granites, near Mt Magnet, and Augrabies, in southern Namibia. The advanced weathering of the fractures, which define the polygonal plates, causes the latter to be reduced in area. Some may be eliminated. Broad flats develop between the surviving plates, which take the form of miniature mesas or mogotes (Twidale, Bourne and Vidal Romaní, 1999). Some slopes, like one on King Rocks, near Hyden in Western Australia, carry many such mogotes (Fig. 11.25).

11.4.2 *Previous interpretations*

Though widely distributed and frequently noted, polygonal cracking has not until recently received the attention it deserves in the geomorphological literature, though it has been attributed to several agencies and mechanisms. Johnson (1927) thought that cracking is due to insolation and chemical weathering. The shells are only thin and within the range of daily and secular temperature changes. But if insolation (or forest fires) has any role in the formation of the cracks, they ought to be preferentially developed on the sunny (in the southern hemisphere, the northern

Figure 11.25. Mogotes developed on King Rocks, east of Hyden, Yilgarn Block, Western Australia.

Figure 11.26. Polygonal cracking on corestones in road cutting in early sixties, Snowy Mountains, New South Wales, 1960.

and western) sides of residual boulders, as well as on exposed upper slopes. They show no such distribution. Moreover, there is some evidence that they are developed on corestones beneath the land surface and at a depth beyond the reach of temperature changes (Fig. 11.26). They could conceivably be related to the ephemeral but intense heat of bushfires, though again depth beneath surface is a problem. Freeze-thaw action involving soil moisture is another possibility in some areas.

Leonard (1929) considered the cracking to be related to the formation of (orthogonal) joints, and though such explanation is germane to those polygonal sets that are developed on structural planes, it cannot apply to those many occurrences on surfaces due to weathering and erosion. Netoff (1971) thought that polygonal cracks in sandstone in Colorado (USA) are due to shrinkage produced by the desiccation of the smectite-bearing bedrock, and this may be valid for the particular occurrence, but the presence of smectite does not appear to be essential to the formation of polygonal cracking. Sosman (1916), on the other hand, concluded that such cracking is due to expansion, to extension of the outer layer of rock. Schulke (1973) identified two types of cracking. He attributed littoral occurrences to the weathering of what he termed exhumed joint planes with which are associated veins of quartz. He did not explain the patterns of fractures, and did not provide evidence for exhumation. In any case, quartz veins are not necessarily or even commonly associated with polygonal cracking, either on the coast or inland. The inland type Schulke (1973) related to the occurrence of duricrusts, their desiccation and cracking as a result of insolation. Robinson and Williams (1989) also favour development on crusted or case-hardened surfaces. The association with case-hardened surfaces is real, but, as argued earlier, cracking is probably not related to insolation, and the encrustation may have developed beneath the land surface, at the weathering front.

11.4.3 *Evidence*

Some features of polygonal cracking are especially characteristic and diagnostic. For example, though the cracking is frequently restricted to the outer shell developed on a boulder or platform, it also occurs on each of several superposed concentric shell layers. The plates are everywhere *in situ*: nowhere have plates been observed displaced (joggled) or collapsed, as would be the case if stretching were involved in their development.

Polygonal cracking is not evenly distributed. Although examples can be found on most outcrops, prolific and well-developed examples are confined to isolated sites. Thus, on northwestern Eyre Peninsula, examples can be located on most of the granite exposures, and well-developed examples have been noted on summit platforms on Corrobinnie Hill and Wallala Hill, but it is only on Tcharkuldu Hill that cracking is abundantly developed and preserved not only on platforms and large residual boulders but also on structural planes. The most critical lines of evidence, however, are first, that spalling is widely developed around boulders, and the resultant shells could form the basis of the eventual plates of polygonal form. Second, weathering produces concentrations of minerals at the weathering front. Third, polygonal cracking can develop beneath the land surface (Fig. 11.26) and fourth, orthogonal patterns of fractures are associated with fault planes and tectonically folded surfaces (Fig. 11.27).

11.4.4 *Explanations*

Polygonal cracking can be divided into two morphological types, each with a separate origin. There are first those sets of cracks, essentially orthogonal in pattern, that are developed on plane, gently arcuate, or warped structural surfaces which are almost certainly planes of dislocation by shearing (Vidal Romaní, 1991). Second there are those many patterns of polygonal cracking that define five- or six-sided plates that are preserved on surfaces due to weathering and erosion, and typically on boulders or platforms.

Orthogonal patterns of cracks are found on essentially plane surfaces in granite, and in sandstone and quartzite. Shearing along the plane of dislocation may have produced not only the orthogonal systems of fractures but also the thin plates on which they are based. That such differential movement has taken place is suggested first by the character of the surfaces which are frequently coated by recrystallised bedrock, and are polished and striated (slickensides). Such orthogonal cracking has been observed also on faults exposed in quarry walls. At some sites, e.g. The Granites, near Mt Magnet, Western Australia, cracking is developed on several layers, and

288   *Landforms and Geology of Granite Terrains*

Figure 11.27.   Polygonal cracking over the outer shell of a dome in Mariz Quarry, Galicia.

Figure 11.28.   Polygonal cracking on several layers at The Granites, Mt Magnet, Western Australia.

some of the vertical cracks can be seen to persist through two or more layers (Fig. 11.28), though others are discontinuous and are confined to a single layer. At Cassia City of Rocks, Idaho, the dislocation surface is undulose and the development of polygonal cracking is patchy, possibly reflecting the discontinuous contact and friction between opposed walls.

Cracking due to shear stresses and consequent changes in planar geometry is well exemplified at the Mariz quarries near Guitiriz, in northwestern Galicia, Spain where an orthogonal pattern of fractures has developed on the crest of a dome (Fig. 11.27).

Turning to the second type of cracking, polygonal patterns are well-developed on boulders and platforms, that is, on surfaces commonly shaped at the weathering front. Thus, at Tcharkuldu, South Australia and The Granites, Western Australia, such patterns are well represented on large residual boulders though at both sites it is also present on planar partings. Insolational heating could cause expansion and cracking. Exposed rock surfaces attain very high temperatures as a result of insolational heating. Rock surface temperatures in excess of 50°C have been recorded in the Egyptian desert and even higher levels (80°C plus) on soils in equatorial Africa, but there is no evidence that ordinary heating and cooling cycles can themselves cause rocks to crack or shatter though rapid cooling associated with desert rain storms may do so: it may be that surface layers expand under the influence of insolational heating and that the stresses caused by rapid cooling induce fracturing at one scale or another.

If such rapid cooling were responsible for polygonal cracking, however, the fractures ought to be most commonly found on upper slopes of boulders and inselbergs. They are, indeed, well-developed on platforms at Corrobinnie Hill, Tcharkuldu Hill and Wallala Hill, on Eyre Peninsula, South Australia but they are at least as common and well-developed on the sides of large residual boulders, and they are also represented on the underslopes of these forms.

Expansion caused by the intense, if ephemeral, heat of bushfires could, in theory, cause the expansion and cracking of the outer shells of rock. Slabs of rock 10–15 cm thick arch from rock masses when they are subjected to intense heating, and it is conceivable that with thin slabs the radial stresses so introduced could result in the development of tangential fractures.

But the field evidence argues against such intense heating playing any part in the development of polygonal cracking. Discontinuous flakes of rock, rather than shells, are produced by bushfires (see Figs 3.2 a and b and Chapter 3). Also, extensive areas of northern Australia and of northern and southern Africa, for example, are deliberately and systematically burned every year or so, and if fires had any part in the development of polygonal cracks, the latter ought to be common in these areas. They are present, but are not notably well represented. They have not been reported from formerly wooded areas such as Dartmoor, southwestern England, which might be expected to have experienced firing. Again, there are reasons for suggesting that the cracking is initiated below the land surface. If this is at all typical, bushfires, and any other form of heating, are ruled out.

Several observations and arguments suggest that polygonal cracking could be initiated beneath the land surface at the weathering front. Examples of polygonal cracking were observed developed on corestones, which are discrete sectors of the weathering front (see Chapter 5), then (early sixties) recently exposed in road cuttings in the Snowy Mountains of New South Wales and in Galicia (Figs 11.26 and 11.27). Concentrations of oxides of iron, manganese and also silica, are typical of the weathering front, and are commonly associated with polygonal cracking on boulders.

The accumulation of minerals at the weathering front could have caused increase in volume, and arching and rupture of the shells in polygonal patterns (Fig. 11.29). This accounts for the observed field data, including the arching of plates as for instance at Tcharkuldu Hill. Repetition of such accumulation and expansion could cause the cracking of lower shells or layers. Stripping of regolith would cause exposure of polygonal cracking and hardening of the mineral-rich shell on drying. In time, however, the indurations are evidently leached out, for there are many examples of older plates (so identified by the wide, weathered cracks between them) which lack discoloration and in fact are bleached, i.e. are paler than the fresh rock (Fig. 11.30). This is not unreasonable, for the ferruginous and manganiferous minerals that are significant components of the surficial indurations are soluble and mobile, and would be leached out after exposure. This suggestion finds support in evidence that polygonal cracking is destroyed by soil moisture for excavations at The Granites (Western Australia) show that even well-developed cracking on exposed surfaces is absent immediately below the surface. Possibly, the soil moisture, rich in organic materials and frequently wetted, is exceptionally efficient in dissolving the indurating salts or in decaying the whole rock.

290   *Landforms and Geology of Granite Terrains*

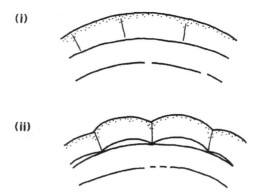

Figure 11.29   Suggested origin of polygonal cracking due to surficial accumulation of oxides of Fe and Mn, and silica (indicated by dots).

Figure 11.30.   (a) Tesselated pavement in granite developed at Boulder Rock, in the Darling Ranges, east of Perth, Western Australia.

Figure 11.30. (b) Tesselated granite surface on the lower slopes of Domboshawa, a bornhardt located a few kilometres northeast of Harare, in Zimbabwe.

11.5 TESSELATED PAVEMENTS

Possibly related to polygonal cracking are the patterns of random cracks which in places produce tesselated pavements in granite. One occurrence can be seen on Boulder Rock, in the Darling Ranges, east of Perth, Western Australia (Fig. 11.30a). There the cracks are indurated with iron oxide derived from a Cretaceous lateritic/bauxitic surface of which remnants persist nearby. Other tesselated pavements have been noted on Domboshawa, a granite bornhardt located a few kilometres northeast of Harare, Zimbabwe (Fig. 11.30b). There the pavement, which shows no obvious iron staining, is clearly exposed from beneath a thin regolith.

REFERENCES

Blank, H.R. 1951. Exfoliation and granite weathering on granite domes in central Texas. *Texas Journal of Science* 3: 376–390.

Boissonnas, J. 1974. Les granites a structures concentriques et quelques autres granites tardifs de la chaîne pan-africaine en Ahaggar, (Sahara Central, Algérie). Centre de Recherches sur les zones arides-Bureau de Recherches Géologiques et Minièrs. Serie Géologie 16. Orleans, France.

Coates, D.R. 1964. Some cases of residual stress effects in engineering work. In Judd W.R. (Ed.), The State of Stress in the Earth's Crust. Elsevier, New York. pp. 679–688.

Ikeda, H. 1996. A strong earthquake and the changes it caused in granitic landscape: the case of Mt Rokko due to the Great Hanshin Earthquake of 1995. *Memoirs of Nara University* 24: 105–116.

Jennings, J.N. and Twidale, C.R. 1971. Origin and implications of the A-tent, a minor granite landform. *Australian Geographical Studies* 9: 41–53.

Johnson, R.J. 1927. Polygonal weathering in igneous and sedimentary rocks. *American Journal of Science* 13: 440–444.

Leonard, R.J. 1929. Polygonal cracking in granite. *American Journal of Science* 18: 487–492.

Netoff, D.I. 1971. Polygonal jointing in sandstone near Boulder, Colorado. *The Mountain Geologist* 8: 17–24.

Plug, C. and Plug, I. 1998. Popups on Moon Rock, Augrabies Falls National Park. *Koedoe* 40(2): 75–84.

Robinson, D.A. and Williams, R.B.G. 1989. Polygonal cracking of sandstone at Fontainebleau, France. *Zeitschrift für Geomorphologie* 33: 59–72.
Rutty, A.L. and Cruden, A.R. 1993. Pop-up structures and the fracture pattern of the Balsam Lake area, southern Ontario. *Géographie Physique et Quaternaire* 47: 379–388.
Schulke, H. 1973. "Schidkrotenmuster" und andere polystrukturen auf felsoberflächen. *Zeitschrift für Geomorphologie* 17: 474–488.
Scott, W.B. 1897. An introduction to Geology. Vol.1 Physical Geology. Macmillan, New York.
Sosman, R.B. 1916. Types of prismatic structure in igneous rocks. *Journal of Geology* 24: 215–234.
Tschang, H.L. 1962. Some geomorphological observations in the region of Tampin, southern Malaya. *Zeitsfricht für Geomorphologie*, 6: 253–259.
Thomas, R.L., Wallach, J.L., McMillan, R.K., Bowlby, J.R., Frape, S., Keyes, D. and Mohajer, A.A. 1993. Recent deformation in the bottom sediments of western and southwestern Lake Ontario and its association with major structures and seismicity. *Géographie Physique et Quaternaire* 47: 325–335.
Twidale, C.R. and Bourne, J.A. 2000. Rock bursts and associated neotectonic forms at Minnipa Hill, northwestern Eyre Peninsula, South Australia. *Environmental and Engineering Geoscience*. 6: 129–140.
Twidale, C.R., Bourne, J.A. and Vidal Romaní, J.R. 1999. Origin of miniature mogotes in granite, King Rocks, southern Yilgarn Block, Western Australia. *Cuaternario y Geomorfología* 13: 33–43.
Twidale, C.R., Campbell, E.M. and Vidal Romaní, J.R. 1993. A-tents from The Granites, near Mt Magnet, Western Australia. *Revue de Gémorphologie Dynamique* 42: 197–203.
Twidale, C.R., Schubert, C. and Campbell, E.M. 1991. Dislodged blocks. *Revue de Géomorphologie Dynamique* 4: 119–129.
Twidale, C.R. and Sved, G. 1978. Minor granite landforms associated with the release of compressive stress. *Australian Geographical Studies* 16: 161–174.
Vidal Romaní, J.R. 1991. Kinds of plane fabric and their relation to the generation of granite forms. *Cuadernos do Laboratorio Xeolóxico de Laxe* 16: 301–312.
Wallach, J.L., Arsalan, A.H., McFall, G.H., Bowlby, J.R., Pearce, M. and McKay, D.A. 1993. Pop-ups as geological indicators of earthquake-prone areas in intraplate eastern North America. In Owen, L.A., Stewart, I. and Vita Finza, C., (Eds), Neotectonic Recent Advances. *Quaternary Proceedings* 3: 67–83.
Whitaker, C.R. 1974. Split boulders. *Australian Geographer* 12: 562–563.
Worth, R.H. 1953. Worth's Dartmoor In Spooner, G.M. and Russell, R.S. (Eds). David and Charles, Newton Abbott. 523 p.

12

Zonality, azonality, and the coastal context

12.1 INTRODUCTION

In geomorphology, zonality and azonality are terms used to denote the distribution of landforms, particularly with respect to global climate, but also in relation to lithology and structure. Zonal forms are those which are restricted to a particular climatic region (or to a particular bedrock), whereas azonal features are not so limited in their distribution.

Several references have been made, in the preceding chapters to forms which are typical of granite landscapes but which are also well-developed in other rock types, and in a range of climatic conditions. Some of these, however, are, apparently, more widely, or are better developed, in some climatic and geological settings than others. No typical granite form is unique, or peculiar, to that rock type and, few, if any, of them are restricted to a single climatic zone (Campbell and Twidale, 1995).

Thus, it was at one time widely held that inselbergs are arid or semi-arid tropical and subtropical forms, and that their occurrence in high latitude regions and in the humid tropics argues climatic change. This trend of thought loses conviction with the realisation (Chapter 6) that many such inselbergs are two-stage forms initiated at the weathering front by the exploitation by shallow groundwaters of structural weaknesses in the country rock, weaknesses which are in many instances of ancient derivation. The regolith is widely developed, and shallow groundwaters are ubiquitous. Thus, such developments are likely to have been widespread, and although proceeding at different rates according to the character of the country rock, and to climatic, and hence shallow regolithic, conditions, the end results are likely to be similar. Differential weathering of the country rock rendered the weaker zones and compartments more susceptible to erosion, mostly by rivers but, according to environment, also by waves, and by glaciers, even by mass movements (solifluction) and by the wind.

12.2 LITHOLOGICAL ZONALITY AND AZONALITY

Some structural features of granite are also developed in other lithological settings and have an overriding influence. Just as regional stress finds expression in fracture patterns and hence landforms in granite, so lineaments have been related to fracture patterns and landform development, for example in the Palaeozoic sandstones of the Grampian Ranges of western Victoria. Sheet fractures and structures are commonplace (Fig. 12.1), as are Kluftkarren, and not only in various climatic settings, but also in coastal as well as terrestrial contexts. In addition, rocks of similar physical qualities tend to produce similar forms. Climate and mineralogy influence the rate of development, but not the end results.

As indicated in Table 12.1 there are no landforms which are exclusively developed on granite. Most forms familiar in granite terrains are also found on various igneous, metamorphic and sedimentary rocks provided the host rock is physically hard and massive, i.e. possessing widely-spaced,

294 *Landforms and Geology of Granite Terrains*

Figure 12.1. Sheet structure in sandstone (a) at Ayers Rock, central Australia, in the Kangaroo Tail, (Wahrhaftig, 1965), (b) in en echelon pattern revealed in a cliff-foot cave at the base of the residual.

clearly-defined and open fractures. Several landforms are not as well or widely developed as on granite, but they are developed in some degree. Thus, bornhardts are a common and characteristic granite form, and they have their equivalents in sandstone and other arenaceous rocks (Young and Young, 1992; Young, 1987). Like their granitic congeners many are developed on fracture-defined blocks (Figs 12.2a and b). Thus, the prominent (200 m high) inselbergs of the Toledo district, Central Spain, e.g. the Cerro Prieto, or the Cerro de Moraján, are outliers of Ordovician quartzite surrounded by Precambrian schist (Fig. 12.2c). Tower-like forms in sandstone are well represented in the Hombori massif of Mali, in the Bungle Bungle massif of the east Kimberleys of Western Australia, in the Monolith Valley of the southern Sydney Basin (inland from Nowra, in New South Wales), in Arnhem Land and Roper River areas of the Northern Territory (St-Onge and McMartin,

Figure 12.1. (c) In sandstone in the Colorado Plateau (Bradley, Hutton and Twidale, 1978); (d) cutting across columnar joints in dacite, Gawler Ranges, South Australia.

1995), and in Ayers Rock, central Australia; and in conglomerate in The Olgas, also in central Australia, in the Meteora complex of central Greece, and in the Mallos de Riglos and Montserrat in the vicinity of the Pyrenees (Fig. 12.3). The Gawler Ranges, South Australia, a massif developed in ignimbritic dacite and rhyolite, is subdivided into large orthogonal or rhomboidal blocks (Fig. 12.2b), and comprising numerous bornhardts, many of them bevelled (Figs 12.3d and e).

The marginal steepening of granitic bornhardts resulting in their conversion to koppies is due to peripheral subsurface weathering, most dramatically manifested in flared slopes; and a similar

Table 12.1. Granitic forms – lithological convergence.

Granite	Other lithologies
Planation Surfaces	
Rolling (peneplains)	
Northern Province, South Africa; northern Eyre Peninsula, SA; southern Yilgarn, WA	Cretaceous sediments – Carpentaria Plains, Qld; Wilcannia area, NSW
Flat (etch or ultiplain)	
Bushmanland, Namaqualand (Western Cape Province), South Africa; Meekatharra, central WA	Sandstone – Bushmanland, Namibia, Namaqualand (Western Cape Province) Schist – central Namibia Limestone – Nullarbor Plain, WA and SA
Pediment – covered	
Central Namibia	Sediments – Flinders Ranges, SA; Cape Fold Belt; American West – Nevada, Utah, Colorado
Pediment – mantled	
Numerous sites	Schist – central Namibia
Pediment – rock	
Corrobinnie Hill, northwestern Eyre Peninsula, SA; many sites in southwestern WA	Argillite – Aliena, Flinders Ranges Conglomerate – west of Olgas, NT Limestone – numerous sites Sandstone – Ayers Rock, NT
Residuals	
Bornhardts	
Numerous sites – either (a) isolated (inselbergs) or (b) components of massifs	Conglomerate – Indonesia; Olgas, NT Dacite/rhyolite – Gawler Ranges, SA Limestone – Indonesia; western Malaysia; southern China (Kweilin) Sandstone – Ayers Rock, NT; Kimba, Eyre Peninsula, SA; Kings Canyon, NT; Mali, West Africa
Nubbins (knolls)	
Many sites in humid tropics (northern and central Australia; Hong Kong) but also Namaqualand (western Cape Province); American Southwest	Dolerite – Keetmanshoop, southern Namibia; western Pilbara, WA
Castle koppies	
Central Zimbabwe; cold or cool uplands in western Europe (southwestern England; Pyrenees; Massif Central; western Iberia)	Sandstone – English Pennines; Roopena, Eyre Peninsula, SA
Towers	
Pitons of Arabia; Guyana; Organ Mountains, New Mexico; Mt Whitney, eastern Sierra Nevada, California; The Needles, South Dakota; Pitoes das Junhas, northern Portugal	Conglomerate – Pyrenees; Meteora, Greece; southern Brazil Limestone – in many areas especially humid tropics – northern Australia: Chillagoe, Qld, Fitzroy Basin, WA; but also Yukon, Canada Sandstone – Gran Sabana, Brazil; Bungle Bungle Ranges, WA; southern Brazil
Pillars	
Murphys Haystacks, Eyre Peninsula, SA	Rhyolitic tuff – City of Rocks, New Mexico Sandstone – Drakensberg, Western Cape Province, South Africa

(Contd)

Table 12.1. (*Contd*)

Granite	Other lithologies
Sheet structure Numerous sites	Dacite – Gawler Ranges, SA Limestone – Arran, Scotland; northern Italy; Appalachians, USA Sandstone – Ayers Rock, NT; Colorado Plateau, USA
Minor Forms *Corestones/boulders* Numerous sites	Basalt – Drakensberg, South Africa Limestone – Galong, NSW Sandstone – Ayers Rock, NT; Flinders Ranges, SA
Pitting Eyre Peninsula, SA; Darwin, NT; NW Qld; WA	Dolerite – Cape Province Limestone – Galong, NSW
Rock basins Numerous sites	Limestone – numerous sites Rhyolite – City of Rocks, New Mexico Sandstone – Ayers Rock, NT; central Spain
Gutters Numerous sites	Sandstone – Drakensberg, S Africa
Grooves Numerous sites	Limestone – Galong, NSW Sandstone – Drakensberg, S Africa
Kluftkarren Numerous sites – Paarl, S Africa; Sierra Nevada, California; Eyre Peninsula, SA	Rhyolite – City of Rocks, New Mexico
Flared Slopes Southern Australia; American Southwest; western and southern France	Basalt – Mexico Dacite – Gawler Ranges, SA Rhyolite – City of Rocks, New Mexico Sandstone – Ayers Rock, NT; Drakensberg, S Africa; English Pennines
Cliff foot caves, notches Podinna, Eyre Peninsula, SA; Devil's Marbles, NT; Balladonia, WA; Kokerbin, WA	Limestone – Sarawak, Malaysia Sandstone – Ayers Rock, NT; Sahara
Scarp foot depressions Central Australia; Wattle Grove and Yarwondutta rocks, Eyre Peninsula, SA	Schist – Egypt Schist & gneiss – eastern Mt Lofty Ranges, SA
Tafoni Numerous sites	Sandstone – NSW; Flinders Ranges, SA; central Spain
Alveoles Point Brown, Eyre Peninsula, SA; Remarkable Rocks, Kangaroo Island, SA; Antarctica	Basic igneous – Point Brown, Eyre Peninsula, SA Sandstone – Beda Valley, SA; Talia, west coast of Eyre Peninsula, SA; Tibooburra, NSW
Rock levées Zimbabwe	Sandstone – Talia, west coast of Eyre Peninsula, SA
Rock doughnuts Central Texas; Kwaterski Rocks, Eyre Peninsula, SA; Galicia, Catalonia, Spain; Zimbabwe	Sandstone – Talia, west coast of Eyre Peninsula, SA; Drakensberg, South Africa

(*Contd*)

Table 12.1. (*Contd*)

Granite	Other lithologies
Minor forms (*Contd*)	
Fonts	
Catalonia, Galicia, and central Spain; Portugal	Basalt – Falkland/Malvinas islands Sandstone – Talia, western Eyre Peninsula, SA; Palmerston, South Island, New Zealand; Cape Paterson, eastern Vic (all coastal); Colorado Plateau, Utah
Plinths	
Eyre Peninsula, SA; Zimbabwe	Sandstone – Talia, west coast of Eyre Peninsula, SA
Pedestal rocks	
Sahara; Appalachians, USA; Sierra de Guadarrama, central Spain	Basalt – Death Valley, California Dolomite – Ciudad Encantada, Spain Sandstone – Kimba, northeastern Eyre Peninsula, SA
A-tents	
Eyre Peninsula, SA; WA; Kulgera, NT	Limestone – NSW Sandstone – Utah, Wyoming, USA
Triangular wedges	
Eyre Peninsula, SA; Mariz, Galicia, Spain; Yilgarn Block, WA	Conglomerate – Baxter Hills, northeastern Eyre Peninsula, SA Dacite/rhyolite – Gawler Ranges, SA
Polygonal cracking	
Eyre Peninsula, SA; Mt Magnet, WA; Galicia, Catalonia, Spain	Sandstone – Flinders Ranges, SA; Litchfield, NT; Fontainebleau, France; Colorado; central Spain
Orthogonal cracking	
Mt Magnet, WA; northern Portugal	Schist – Galicia, northwestern Spain
Speleothems	
Eyre Peninsula, SA; Kulgera, NT; Mt Magnet, WA; Galicia, Spain	Conglomerate – Baxter Hills, northeastern Eyre Peninsula, SA Limestone – numerous sites Sandstone – Darwin area, NT; Flinders Ranges, SA

States of Australia: NSW – New South Wales, NT – Northern Territory, Qld – Queensland, SA – South Australia, Vic – Victoria, WA – Western Australia; USA – United States of America.

mechanism appears to be at work in sandstone terrains, with the development of cliff-foot caves and shelters, as for example at Ayers Rock (Fig. 12.4a). Cupolakarst is similar, and is also converted to towerkarst or turmkarst (Fig. 12.4b) as a result of marginal weathering and steepening. Some of this solution takes place beneath the soil surface but also finds surface expression in swamp slots (Fig. 12.4c).

Many, perhaps most, granitic inselbergs appear to be two-stage forms based on massive compartments. But some are an expression of lithology (Chapter 6) and many non-granitic residuals are of this type. They are upstanding by virtue of their development in resistant rock. Thus, The Pinnacles, near Broken Hill, in western New South Wales (Fig. 12.5a) are upstanding by virtue of the presence of discrete masses of quartzite which confer a measure of resistance on the otherwise weak schists into which they are injected and in which the adjacent plains are eroded. Some well-known isolated steep-sided hills, such as the Glasshouse Mountains of southeastern Queensland, Ship Rock, New Mexico (Mueller and Twidale, 1988), and Wase Rock, northern Nigeria, are volcanic plugs (Fig. 12.5b). Again, Curtinye and Barna hills are low quartzitic domes near Kimba on northeastern Eyre Peninsula, South Australia. Their domical form reflects the development of rudimentary sheet fractures which cut across a steeply-dipping foliation, but the resistance of these

Zonality, azonality, and the coastal context 299

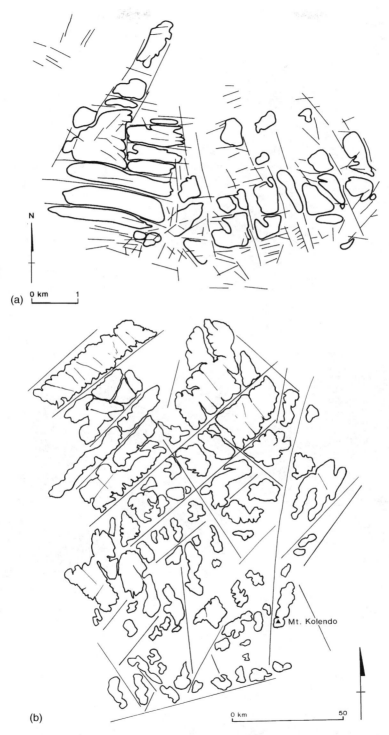

Figure 12.2. (a) Plan of the Olgas complex, central Australia, showing relationship between fracture pattern and residuals. (b) Plan of part of the Gawler Ranges, South Australia, showing relationship between fracture pattern and residuals.

Figure 12.2. (c) Inselberg of Cerro de Moraján, Toledo, central Spain developed on Ordovician quartzite standing in Precambrian schist.

Figure 12.3. (a) Ayers Rock, a bevelled bornhardt developed in arkosic sandstone, in central Australia. Stratigraphic evidence points to the upper surface being of etch origin and of later Cretaceous age. (S.A. Tourist Bureau).

particular compartments and the development of the sheeting due to kinks, or locally developed tight folds, and hence to compressed sections, in the metamorphic sequence.

Nubbins are well-developed in granite and also in amphibolite, dolerite and norite (Hutton, Lindsay and Twidale, 1977) where they are related to orthogonal fracture systems. They are also

Zonality, azonality, and the coastal context 301

Figure 12.3. (b) Conglomeratic towers of the Olgas complex, central Australia, and (c) at Meteora, in central Greece. Note the flared sidewalls. (d) Bevelled domes in dacite in the southern Gawler Ranges, South Australia.

302 *Landforms and Geology of Granite Terrains*

Figure 12.3. (e) Bevelled domes in dacite in the southern Gawler Ranges, South Australia.

Figure 12.4. (a) Cliff-foot cave at the base of Ayers Rock, central Australia.

found on sandstone where strata have broken down into large blocks and slabs. Koppies, on the other hand, are well-developed in sandstone, for example in the English Pennines, where basal fretting is also evident. Rare examples are also found in dacite, in the Gawler Ranges (Fig. 12.5c). All slopes topography (Chapter 7) is well represented in argillites in the Pertnjara Hills, southwest of Alice Springs, Northern Territory (Fig. 12.6).

Boulders are commonplace in granitic terrains but two-stage (multi-stage) features are also found in limestone (Karrensteine, Karrenblöcke) and in sandstone, as well as in igneous rocks such as basalt, dacite, norite, etc. (Fig. 12.7). They are associated with orthogonal fracture systems (Fig. 12.8a) and like their granitic congeners display corestone boulders with flaking and spalling (Figs 12.7a and 12.8b).

All the minor forms listed and discussed in the various relevant chapters have their analogues and congeners in other rock types. Even pitting has its equivalents (Fig. 12.9). This is to be expected not only of those several features that are of etch origin, but also of those associated with

Zonality, azonality, and the coastal context 303

Figure 12.4. (b) Cupolakarst (mogotes) in Puerto Rico (W.H. Monroe). (c) Swamp slot in limestone at Ipoh, West Malaysia.

tectonic stress, such as A-tents, for tectonism, like the regolith and groundwaters, is ubiquitous, though more prevalent in some areas than others. But it is not surprising that A-tents have been reported developed not only in granite but also in similarly brittle materials like crystalline limestone and well cemented (quartzitic) sandstone, and they have been reported in sandstone from Wyoming, in limestone from the Coolamon Plain, New South Wales, and in marble at Wombeyan Caves, also in New South Wales. Likewise, dislodged blocks are found in sandstone as in the Roraima Plateau, Venezuela, and dislodged or collapsed columns are found in the Gawler Ranges in dacitic welded tuffs.

Flared slopes are found well-developed not only on granite but also on sandstone and conglomerate, for example at Ayers Rock and the Olgas, in the English Pennines and in the Drakensberg of southern Africa (Figs 12.9a and b). They occur in trachyte at Hanging Rock,

304 *Landforms and Geology of Granite Terrains*

Figure 12.5. (a) The Pinnacles, western New South Wales. (b) Wase Rock, a plug of Tertiary trachyte in northern Nigeria (Aerofilms, London). (c) Small koppie in dacite, Gawler Ranges, South Australia.

Zonality, azonality, and the coastal context 305

Figure 12.6. All slopes topography in argillites, Pertnjara Hills, southwest of Alice Springs, Northern Territory (Wahrhaftig, 1965).

Figure 12.7. Corestone boulders (a) in basalt, southern Drakensberg, Cape Province, South Africa.

Victoria (Spencer-Jones, 1965), and, albeit poorly developed, in dacite in the southern Gawler Ranges. But they occur in abundance in a rhyolitic tuff in the City of Rocks, Grant County, southern New Mexico. Here the walls of the fracture-defined corridors (or streets) are all to a greater or lesser degree flared; and the subsurface origin of the forms is demonstrated in several minor excavations due to accelerated soil erosion in the floors of the corridors (Figs 12.9c and d).

Several other minor forms typical of granite are developed in sandstone. Huge rock basins and doughnuts occur in the Colorado Plateau, in the Utah-Arizona border region (Bradley, 1963; Netoff, Cooper and Shroba, 1995 and see Figs 12.11a and b). They are also developed in arenaceous

306 *Landforms and Geology of Granite Terrains*

Figure 12.7. (b) In sandstone, southern Flinders Ranges, South Australia.

Figure 12.8. (a) Orthogonal fractures in dolerite, near Umtata, Transkei, South Africa.

rocks in other areas. For instance, doughnuts and pedestal rocks occur in the southern foothills of the Drakensberg (Figs 12.11c and d). One rudimentary and irregular, but nevertheless definite, rock doughnut has been noted in dacite in the Scrubby Peak region of the southern Gawler Ranges. Rock basins are well-developed both in limestone, where they are known as kamenitzas, and in sandstone, as on Ayers Rock, in the southern Flinders Ranges, South Australia, and in the southern Drakensberg. Tafoni are well-developed on basaltic and trachytic exposures, as in Andén Verde, Gran Canaria, Canary Islands (Fig. 12.12), and are also found on sandstone boulders and at the base of bluffs. Alveoles, though developed on granite surfaces, are at many sites better developed on such materials as sandstone, basalt, amphibolite and siltstone, both at inland and in

Figure 12.8. (b) Corestone-boulder of norite, and with spalled margins, Black Hill, western Murray Basin, South Australia.

Figure 12.9. (a) Intricate differential weathering at crystal scale in limestone, Galong Quarry, central New South Wales.

coastal settings see page 313. Like their granitic counterparts, some flutings in limestone originate beneath the land surface (Fig. 12.9a), as do gutters in sandstone (Fig. 12.13). Similarly, caves are synonymous with carbonate and gypseous formations, but, though present in a few granitic masses, they are not so common and well-developed. Though well represented in granitic terrains, siliceous speleothems are also widely developed in some sandstone settings, e.g. the Flinders

Figure 12.9. (b) Fretted pattern on surface of dolerite corestone, developed immediately below land surface, but now released and exposed in roadworks, near Umtata, Transkei, South Africa.

Figure 12.10. Flared slopes in sandstone (a) southern base of Ayers Rock, central Australia, (b) Drakensberg of South Africa.

Ranges northeastern Eyre Peninsula, South Australia, Litchfield National Park, near Darwin, Northern Territory and in conglomerate in the Baxter Hills, northeastern Eyre Peninsula, South Australia.

Mantled and rock pediments are well-developed in other massive rocks, as for example around Ayers Rock and west of The Olgas, where an extensive plain in conglomerate is in part an exposed platform, in part a mantled pediment.

Zonality, azonality, and the coastal context 309

Figure 12.10. (c) Subsurface initiation of flares at City of Rocks, exposed in a shallow gully (Courtesy of Mueller). (d) A street with flared sidewalls developed in rhyolitic tuff, City of Rocks, New Mexico (Courtesy of Mueller).

Figure 12.11. (a) Large rock basins.

12.3 CLIMATIC ZONALITY AND AZONALITY

Several well-known granite forms like bornhardts, boulders and basins are climatically azonal. They are shaped by different, climatically-driven processes, but the end results are the same or similar: convergent forms.

Figure 12.11. (b) Conical fonts (30 m high) in sandstone, Colorado Plateau, Utah. (c) Doughnuts in sandstone.

Other landforms, however, have restricted distributions in terms of climate. Thus, tafoni are well-developed in arid and semi-arid lands, hot and cold, and also in arid and semi-arid coastal zones, so that they are well-developed in Antarctica, in mediterranean lands like Catalonia, northeastern Spain, Corsica and Eyre Peninsula, as well as in the hot deserts of Namibia and central Australia. Tafoni are also well-known from humid interior lands, though whether this reflects an

Zonality, azonality, and the coastal context 311

Figure 12.11. (d) Pedestal rock in sandstone, southern Drakensberg, eastern Cape Province, South Africa.

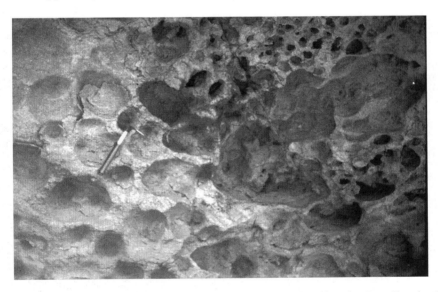

Figure 12.12. Tafone with honeycomb developed in trachyte from Andén Verde, Gran Canaria, Canary Islands, Spain.

absence of salt weathering or its being overshadowed by other, more rapidly acting, processes such as hydration and hydrolysis causing the destruction, e.g. of the visor, is not clear. Tafoni are well-developed in Galicia, not only in humid coastal areas but also inland, where hydration and hydrolysis may assume a greater role in their development.

So strong is the link between climate and landform thought to be in some instances, that climatic changes have been predicated on the basis of the relationship (Twidale and Lageat, 1994). Thus, Granitrillen, developed and preserved in the Karkonosze uplands of southwestern Poland, have been attributed to Tertiary humid climates, on the assumption that they could not have formed in moderately humid or periglacial climates (Migon and Dach, 1995), though this deduction is called

312 *Landforms and Geology of Granite Terrains*

Figure 12.13. Decantation gutters in sandstone, southern Drakensberg, Eastern Cape Province, South Africa. The gutters have been exposed by the recession of the edge of the thin regolith.

Figure 12.14. Towers in granite, The Needles, South Dakota.

into question by the late Holocene development of similar forms on prehistoric human monuments in Brittany (Fig. 8.23).

Nubbins are prominent in humid tropical regions, such as monsoonal northern Australia and Hong Kong, and this can be explained in terms of the rapid disintegration of the outer shells or sheet structures of bornhardts in the high temperatures and seasonally abundant groundwaters typical of such regions. Sheet structure is weathered and broken into blocks beneath ground level. Nubbins are however known from other climatic settings such as Namaqualand, in Western Cape Province of South Africa, and the Alice Springs area of the Northern Territory, Australia. In each of these examples, the nubbins occur in topographic basins to which surface and groundwaters gravitate so that they may develop in local moist sites. On the other hand, nubbins in present mid-latitude cool areas, such as the Bohemian Massif (Demek, 1964), and desert settings, as in the Mojave Desert of the southwestern United States (Oberlander, 1972), have been attributed to development in warm humid climates followed by climatic change.

Castle koppies have a peculiar bimodal distribution, or apparently so. On the one hand, they are characteristic of cold lands, and are the well-known tors of the granite uplands of southwestern England (Linton, 1964). They are also well represented in the Pyrenees, the Bohemian Massif and the Massif Central of France. They appear to be due to intense frost action and to ground-ice active in the regolith at the margins of the residuals. On the other hand, the classical koppie country is the high veld of southern Africa (Lageat, 1989) and especially Zimbabwe. They are associated with palaeosurfaces of low relief. Here the koppies appear to be the result of long-continued subsurface marginal weathering or residual masses the crests of which are just exposed. As with nubbins, koppies are also developed in local wet sites, as at the Devil's Marbles, Northern Territory (Fig. 7.9). Thus, the basic mechanism for koppie development may be the same though the processes involved are very different. In these terms koppies are convergent landforms.

Frost action is involved in the formation of another distinctive residual in granite. The freeze-thaw mechanism is evidently effective in exploiting well-developed fractures, and for this reason towers and frost-shattered bedrock are well represented in cool or seasonally cool climates. Good examples are found in the Organ Mountains of southern New Mexico, in Northern Canada, in the Mt Whitney area of the eastern Sierra Nevada, and in Cathedral Rocks, in the Yosemite, both in California (Figs 7.12b and 7.13). The Needles, in South Dakota, provides another well-known example (Fig. 12.13). The twin peaks of Fitz Roy and Cerro Torre in the Province of Santa Cruz, in Argentine Patagonia, and, although smaller, several hills in the Sierra de Gredos of the Spanish Central Massif and in the Pitões das Junhas of northern Portugal, provide other examples.

Though sheet structure is widely developed, pseudobedding (Fig. 2.2) restricted to near surface zones may be due to frost action, to the freezing and expansion of meteoric waters that percolated into the country rock, causing the separation of one layer after another. Such features may well, however, reflect exploitation by weathering agencies of latent stress fractures or lineations in the rock.

12.4 THE COASTAL CONTEXT

Many typical granite forms are found in coastal zones in various parts of the world. Some are structural in origin. Thus, the sheet structures exposed in the littoral of the Pearson Islands (Fig. 1.8), in the eastern Great Australian Bight, are most likely (see Chapter 2) a manifestation of tectonic stress, and are identical with those exposed, for example, in the sidewalls of deeply incised valleys (Fig. 2.6). Again, geos due to the exploitation of fracture zones, or, more commonly, to the preferential weathering and erosion of weaker veins intrusive into the granite host mass, are fairly common, and sea caves, characteristically pear-shaped in cross section (Sjoberg, 1986, 1987 and see Fig. 12.15), due to the exploitation by waves of steeply inclined fracture zones have been reported from several sites on the Swedish coast, and inland, where they are used as indicators of isostatic movements (Sjoberg, 1986, 1987).

314 *Landforms and Geology of Granite Terrains*

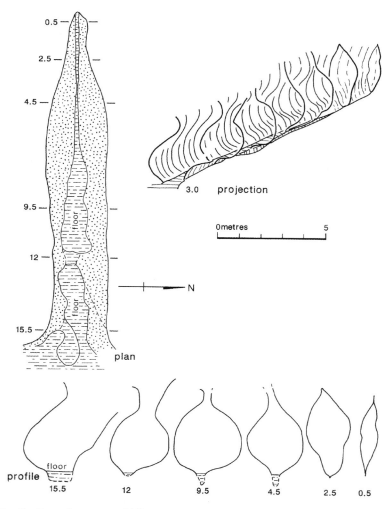

Figure 12.15. Sections of sea cave at Lidbergsgrottorna, northern Sweden.

Other forms are inherited from the adjacent land mass: they have been developed by processes active in the terrestrial context, but have been exposed at the coast and by marine agencies. Thus, granite corestones set in a matrix of grus are exposed in low cliffs west of Windmill Bay, near Cape Willoughby, on the southeast coast of Kangaroo Island, and in front of the cliff, the beach consists of granite corestones released from the regolith by wave erosion of the matrix (Fig. 12.16a). Some of the released corestones differ from their terrestrial counterparts in that, like boulders in other lithologies exposed on high energy coasts, they are faceted, scalloped, polished and fitted (Hills, 1970) as a result of grinding induced by powerful storm waves (Fig. 12.16b), but otherwise they are comparable to any other corestone boulders. The boulders spread over the platforms at Smooth Pool, on the west coast of Eyre Peninsula (Twidale, Bourne and Twidale, 1977 and see Fig. 12.17a), and at various sites on the Galician Atlantic coast, are also released corestones, for remnants of a grussy matrix and regolith are preserved in places, and a few of the boulders display flared sidewalls (Fig. 12.17b).

Again, literal and littoral inselbergs occur in coastal zones in several parts of the world. Some, like those near Esperance in Western Australia and Rio de Janeiro, in southeastern Brazil, are

Zonality, azonality, and the coastal context 315

Figure 12.16. (a) Boulder or shingle beach at Windmill Bay, Kangaroo Island, South Australia, composed of released granite corestones. (b) Fitted and polished boulders at Windmill Bay.

domical, but off Cape Wilson, southern Victoria, both nubbins and bornhardts form islands (Fig. 12.18). Some are probably upstanding by virtue of their low fracture density, like Remarkable Rocks on the southwestern coast of Kangaroo Island, but others, like those of Encounter Bay, in South Australia, are an expression of lithological contrast, for the granite masses forming islands and headlands are intrusive into schists and gneisses where have been preferentially weathered and eroded by epigene or marine agencies, or by both. Flutings exposed on granite blocks near Hillock Point, on southern Yorke Peninsula, South Australia, predate the dune calcarenite from beneath which they are being exhumed, and may be of epigene provenance. Minor Kluftkarren exposed in the coastal zone at Point Brown, west coast of Eyre Peninsula, give rise to an assemblage comparable

316   *Landforms and Geology of Granite Terrains*

Figure 12.17. (a) Shore platform in granite at Smooth Pool, near Streaky Bay, west coast of Eyre Peninsula, South Australia, with released corestones scattered over surface, and (b) released corestone with flared sidewall, Smooth Pool.

to the clint and grike of karst pavements, and are most likely inherited from terrestrial weathering and exploitation for the clefts are flared (Fig. 12.19a). On the other hand minor but numerous fracture-controlled linear grooves at Cape Willoughby, eastern Kangaroo Island, and Arteixo, A Coruña, northwestern Spain, are opened and cleaned by spray and waves (Figs 12.19b and c), though as at Point Brown, remnants of a flaked or laminated bedrock are preserved between blocks and boulders, and a regolith is widely preserved beneath the dune calcarenite, confirming the suggestion of terrestrial preparation and an etch origin (Fig. 12.19d).

Several landforms are common to exposures in both the terrestrial and coastal contexts. Some coastal representatives have developed in a manner similar to their terrestrial congeners. Thus, the doughnuts and fonts of the Talia coast of Eyre Peninsula, South Australia (Figs 12.20 a–c) appear to be similar in origin to doughnuts in granite from inland sites, but here it is beach sand rather

Zonality, azonality, and the coastal context 317

Figure 12.18. (a) Nubbin and (b) bornhardt forming islands in the Glennie Group, at Wilsons Promontory, Victoria.

than the regolith that retains the moisture that produces rapid weathering of the surrounding areas. The fonts of Shag Point, near Moeraki, South Island, New Zealand (Fig. 12.20d), on the other hand, are indurated with limonite; their evolution is not yet known. The runnels found on some limestone ramps on the west coast of Eyre Peninsula in a zone seaward of an inner zone of potholed ramp or platform are, like their terrestrial counterparts, sculpted by running water, but here by the wash and backwash of breaking waves.

Other forms common to terrestrial and coastal sites have, however, evolved in different ways and are thus convergent forms. Plinths provide a good example. As described in Chapter 9, plinths on inland granitic residuals are due to the umbrella effect of the block or boulder that rests or

Figure 12.19. (a) Kluftkarren, or clint and grike in granite, Point Brown, northwestern Eyre Peninsula, South Australia. Clefts (grikes), about 15 cm deep. (b) Shallow fracture-controlled clefts, Cape Willoughby, Kangaroo Island, South Australia. (c) Kluftkarren on a vertical wall in leucogranite from Arteixo, A Coruña, Galicia.

rested upon them. This is demonstrated by the correlation in plan shapes of the protective residual and of the plinth. Similar features are common on the Talia coast of western Eyre Peninsula, but here the plinths are of sandstone and the blocks resting on them consist of dune calcarenite (some with calcrete) fallen from the adjacent cliffs. The plinths are due not so much to the protection

Zonality, azonality, and the coastal context 319

Figure 12.19. (d) Regolith (X-X) developed on granite and preserved beneath dune calcarenite, Point Brown, northwestern Eyre Peninsula, South Australia. Note lamination at base of regolith.

Figure 12.20. Minor forms developed on shore platform in sandstone, Talia, west coast of Eyre Peninsula, South Australia: (a) Doughnut (or shirt collar). Note the plinth with calcarenite block in middle distance.

Figure 12.20. (b) Steep-sided tower-like fonts; (c) conical font.

afforded by the limestone blocks but rather to the increased turbulence and diversion of waves by the blocks. In this way, the sandstone on which a block rests facing the sea is preferentially scoured, forming a low cliff and the beginnings of a plinth. Similarly, waves diverted around the sides of the block erode the platform there, again leaving the base on which the blocks rest in relief, as a plinth (Fig. 12.21a).

Many rock basins (overwhelmingly of the hemispherical or pit type – see Chapter 9) are pot-holes due to abrasion by cobbles and small boulders swirled around by waves. They have their congeners in granitic stream channels though the basins of terrestrial granite outcrops are due predominantly

Zonality, azonality, and the coastal context 321

Figure 12.20. (d) Conical font in sandstone, Shag Point, Moeraki, South Island, New Zealand.

Figure 12.21. (a) Sandstone plinth with calcarenite block, Talia, west coast of Eyre Peninsula, South Australia.

to chemical attack of standing water. At most sites, only granite sand is available, and in any case only on lower slopes is runoff of sufficient velocity and volume to set the sand in motion and use it as an abrasive tool. Only in river and subglacial stream beds, and adjacent to marine cliffs, where breaking waves induce turbulence and grinding action, has pot-holing contributed significantly to the development of basins in granite bedrock, and many of these are more closely allied to the cylindrical hollows described in Chapter 9. Some forms are comparable to terrestrial counterparts, but are especially well and widely developed because of the oceanic or coastal environment. Thus, the abundance of alveoles (developed in sandstone (Fig. 12.21b), mudstone and amphibolite as well as granite) and tafoni reflect the ready availability of salt, which, especially in dry or seasonally dry climates, can precipitate out and shatter rocks.

322   *Landforms and Geology of Granite Terrains*

Figure 12.21.   (b) Alveoles in sandstone exposed in cliff at Talia, west coast of Eyre Peninsula, South Australia.

Others are peculiarly coastal. Thus, shore platforms are obviously not found elsewhere, though whether they are characteristic of granite exposures has been disputed. In Victoria, Hills (1949) found no platforms developed on granite, though they are more or less well-developed on many other rock types. On the other hand, shore platforms are well-developed in granite on the west coast of Eyre Peninsula, and at other Australian sites, e.g. Coles Bay in Tasmania. In some areas, and especially on the west coast of Eyre Peninsula, the platforms are of etch origin, being due to the stripping by waves and other marine agencies, of a regolith, remnants of which are widely preserved beneath a late Pleistocene dune calcarenite (Fig.12.19d). Again, shore platforms of this kind are well-developed in granite on the coast of Galicia, and also in western Canada where they are part of the strandflat (Fig. 12.22). In part, this mild difference of perception and interpretation may be a semantic problem, in that there is a gradation between quite smooth platforms (like that at Smooth Pool, northwestern Eyre Peninsula) and irregular, blocky and bouldery, yet overall quite gently inclined, features like that exposed at Point Drummond, southwestern Eyre Peninsula (Fig. 12. 23). Fitted blocks have already been mentioned, and pecking, the apparent chipping out

Figure 12.22. (a) Shore platform in granite, Costa da Morte, Galicia, Spain. (b) Granitic sector of strandflat, coast of northern British Columbia, Canada (Department of Energy and Natural Resources, Canada).

Figure 12.23. Irregular shore platform at Point Drummond, west coast of Eyre Peninsula, South Australia, produced by the stripping of the 20 m thick regolith preserved in the promontory: an etch form.

of shallow angular depressions a few centimetres diameter and deep is typical of some granite (and sandstone) coasts. It is not due to physical abrasion, to the impact of cobbles and blocks carried by the waves, for there is no distributional relationship between the pecking and either exposure to wave attack or to availability of coarse debris.

REFERENCES

Bradley, W.C. 1963. Large-scale exfoliation in massive sandstones of the Colorado Plateau. *Geological Society of America Bulletin* 74: 519–528.
Campbell, E.M. and Twidale, C.R. 1995. Lithological and climatic convergence in granitic morphology. *Cadernos Laboratorio Xeoloxico de Laxe* 20: 381–403.
Demek, J. 1964. Castle koppies and tors in the Bohemian Highland (Czechoslovakia). *Biulytyn Periglacjalny* 14: 195–216.
Hills, E.S. 1949. Shore platforms. *Geological Magazine* 86: 137–152.
Hills, E.S. 1970. Fitting, fretting and imprisoned boulders. *Nature* 226 (5243): 345–347.
Hutton, J.T., Lindsay, D.M. and Twidale, C.R. 1977. The weathering of norite at Black Hill, South Australia. *Journal of the Geological Society of Australia* 24: 37–50.
Lageat, Y. 1989. Le Relief du Bushveld. Une Géomorphologie des Roches Basiques et Ultrabasiques. Faculté des Lettres et Sciences Humaines de L'Université Blaise-Pascal, Nouvelle Série, Fascicule 30.
Linton, D.L. 1964. The origin of the Pennine tors – An essay in analysis. *Zeitschrift für Geomorphologie* 8: 5–24.
Migon, P. and Dach, W. 1995. Rillenkarren on granite outcrops, SW Poland, age and significance. *Geografiska Annaler* 77A: 1–9.
Mueller, J.E. and Twidale, C.R. 1988. Geomorphic development of City of Rocks, Grant County, New Mexico. *New Mexico Geologist* 10: 74–79.
Netoff, D.I., Cooper, B.J. and Shroba, R.R. 1995. Giant sandstone weathering pits near Cookie Jar Butte, southeastern Utah. In C. van Rieper III (Ed.), Proceedings of the Second Biennial Conference on Research in Colorado Plateau National Parks (Flagstaff, Arizona, October, 1993). United States Department of the Interior, National Parks Service. pp. 25–53.
Oberlander, T. 1972. Morphogenesis of granitic boulder slopes in the Mojave Desert, California. *Journal of Geology* 80: 1–20.

St-Onge, D.A. and McMartin, I. 1995. Quaternary Geology of the Imman River Area, Northwest Territories. Geological Survey of Canada Bulletin 446.

Sjoberg, R. 1986. Tunnelgrottor i Norden – geomorfogiska studier. Gerum Naturgeografi (Geografiska Institutionen Umeå Universitat) 3.

Sjoberg, R. 1987. Caves as indicators of neotectonics in Sweden. *Zeitschrift für Geomorphologie* SupplementBand 63: 141–148.

Spencer-Jones, D. 1965. The geology and structure of the Grampians area, western Victoria. Memoirs of the Geological Survey of Victoria 25.

Twidale, C.R., Bourne, J.A. and Twidale, N. 1977. Shore platforms and sealevel changes in the Gulfs region of South Australia. *Transactions of the Royal Society of South Australia* 101: 63–74.

Twidale, C.R. and Lageat, Y. 1994. Climatic geomorphology – a critique. *Progress in Physical Geography* 18: 319–334.

Young, R.W. 1987. Sandstone landforms of the tropical East Kimberley region, north-western Australia. *Journal of Geology* 95: 205–218.

Young, R.W. and Young, A. 1992. Sandstone landforms. Springer, Berlin.

13

Retrospect and prospect

"Ages are spent collecting materials, ages more in separating and combining them. Even when a system has been formed, there is still something to add, to alter or to reject. Every generation enjoys the use of a vast hoard bequeathed to it by antiquity, and transmits that hoard, augmented by fresh acquisitions, to future ages. In these pursuits, therefore, the first speculators be under great disadvantages, and when they fail, be entitled to praise." (Lord Macaulay, Essay on Milton).

No landform is restricted to granite. Many are better or more commonly developed in granite than in any other rock type but none is unique to that material. All have their congeners in other lithological environments. The corollary of this is that the factors, processes and mechanisms invoked in explanation of granite landforms have a wider application and indeed are germane to the understanding of landforms and landscapes in general. Thus, many granitic forms are of two-stage origin, and originated at the weathering front. There, shallow subsurface groundwaters exploited structural weaknesses, major and minor. There can be no doubt that many familiar landforms have their origins in the subsurface though it would be imprudent to assume that all representatives of a particular form are of such an origin. There can be no doubt, for example, that some runnels or flutings, and some rock basins, have formed after the exposure of the host surface, though equally, many demonstrably have formed at the weathering front. Similarly, though many bornhardts are of two-stage origin, others, due to upfaulting, for example, may well be epigene features. The local evidence must be sought. If the problem is appreciated, however, the greater the chance of finding relevant evidence.

At a quite different scale, pitting, notches and flared slopes are good indicators of recent soil erosion and of the thickness of soil evacuated (e.g. Fig. 13.1a; also Fig. 3.5e). The flared, and notched, basal slopes shown in Fig. 13.1b, suggest that the thickness of detritus stored in these basins and valleys on the flanks of Domboshawa, a bornhardt near Harare in central Zimbabwe, was formerly much greater.

The implications of such a two-stage origin are several, but in particular, besides emphasising the great antiquity of the geological factors in part responsible for many contemporary forms, there is a strong possibility that as groundwaters are ubiquitous, so are the forms they generate. Many granitic forms, and their congeners in other rock types, appear to be azonal. This is a conclusion of general significance.

The Earth's crust is not quiescent but continues to be active – more at and near plate margins than elsewhere but everywhere active. Differential movements cause stress and strain, as witness the many modern or neotectonic landforms known from both granitic and other terrains, and manifested also in contrasted stress regimes and in varied fracture densities. These crustal characteristics find particularly clear expression in granitic terrains.

At and near the Earth's surface granite is brittle and neotectonic forms, involving warps as well as fractures, are well-developed and preserved. Minor scarps, A-tents and wedges are especially well represented and preserved, but antiforms and synforms also find topographic expression. Tectonic stress is significant in the formation of such features as bornhardts and sheet structure but

Figure 13.1 (a) Flared notch, on granite slope in Namaqualand (Western Cape Province), South Africa indicating former presence of soil cover. (b) Part of the upper slope of Domboshawa, a bornhardt near Harare in central Zimbabwe. The basin and valley (on left) are occupied by detritus, but that this fill was once much greater is indicated by the flared slopes (formed beneath a soil cover) and by the notch etched into the base of the slope in the depression nearest the camera.

it is not suggested that bornhardts for instance are always tectonic forms: some are, yet others have their origins in crustal events. Tectonic stresses have paved the way for later developments by, for example, determining fracture spacing and condition.

Other stresses however have played their part and ought not to be neglected. Gravitational loading and tectonic pressure, i.e. the pressure exerted by one mass on the one against which it is being pressed, either through gravity or by crustal stress, may cause strain or distortion in mineral texture and thus render the rocks more susceptible to weathering and hence erosion. Such loading may explain on the one hand the location of basins and tafoni, and on the other aligned basins and gutters at sites where no fracture is apparent.

In this survey and analysis of the landforms developed in granitic terrains, the part played by interactions between country rock and water, and hence the critical importance of rock structure in facilitating – or inhibiting – water infiltration, has been emphasised. The role of water as a transporting medium has also been highlighted. Water takes not only chemicals derived from the atmosphere and the regolith but also biota, especially bacteria, into the rock mass. It translocates and evacuates the products of alteration resulting from reactions between the rock-forming minerals and water charged with chemicals and biota. In suitable environments and circumstances it carries solids in the subsurface, in some areas sufficiently to cause surface subsidence as a result of volume loss and compaction.

Water is supremely important in shaping the Earth's surface. Its effects are, however, more clearly demonstrated in granitic terrains than in most. Dry granite is stable but in contact with water rapidly decays. When fresh, granite, being crystalline, is of low porosity and permeability, but it is also characteristically pervious so that water infiltration is concentrated along partings. Thus, fracture density and geometry substantially determine topographic patterns at a range of scales from inselbergs and plains, bornhardts and boulders, basins and gutters. Fracture geometry, in turn, reflects not only the composition and texture of the granite but also the strength and sense of crustal stresses, and the environment (pressure, temperature, and hence rheology) of the affected granite at the time of stress.

The part played by water in shaping the land surface underlines not only the importance of avenues of entry and weathering such as fractures, but also the general point that dry sites may survive whereas moist ones tend to decay. And this in turn points up the importance of temporary water retention, in regoliths, for example, in landform and landscape development. Uplands which shed water tend to remain upstanding whereas lowlands are repeatedly weathered, eroded and lowered. Reinforcement or positive feedback mechanisms are as important in granitic terrains as they are elsewhere, and at various scales.

Most of the changes effected by the contact of water with granite take place beneath the land surface, at the base of, or within, the regolith. At the weathering front groundwaters exploit variations (weaknesses) in the bedrock with which they are in contact. Some of those variations have their origins in magmatic, thermal and dislocational events that took place long ago, even in geological terms. Thus, in a very real sense the origin of many landforms can be traced back tens or hundreds of millions of years, or even a billion or more years. Many landforms are really multi-stage in that sense, as well as varying in their development after exposure. The two stages of development frequently cited, the preparation of forms at the weathering front, and their subsequent exposure, are, however, critical.

The changes resulting from the interaction of granite and water do not take place instantaneously, but in time, and the longer the contact, the greater the degree, say, of the rounding of a fracture-defined block and its conversion to a corestone/boulder. That is why, on the one hand, many inselberg landscapes are associated with long exposed shield areas, and why on the other, and in general terms, the shallower the depth below the surface, the more advanced are weathering and landform preparation.

Atmospheric climate obviously affects the physics and chemistry of groundwaters, both directly (e.g. temperature) and indirectly (e.g. organic acids, biota), and it thus probably affects the rate at which granite landforms develop. And there are some forms which appear to be zonal, that is they are found mainly in specific climatic regions. Nubbins (humid tropics) and tafoni (arid or semiarid areas) are two that come to mind. Granite areas affected by frost and ice are superficially shattered, and angular peaks and screes for example are prominent, but so are similar features on other massive bedrock exposures, and many features typical of granite in temperate or tropical regions are also found in cold areas.

On the other hand, many granite landforms, major and minor, are azonal. This results in part from the control exerted by structure *sensu lato*, but is also due to many of them being of etch origin. Shallow groundwaters are ubiquitous, so that differential weathering at the base of the regolith, at the weathering front, is also widespread. Though the rate and precise character of weathering processes, and in some degree erosion, varies with atmospheric and regolithic climate, the end results are similar. Also, and for the same reason, many landforms characteristic of granite are also found well-developed on rocks that are genetically different from, but physically similar to, granite.

The study of granitic landforms and terrains well demonstrates the relationships between major and minor forms. Thus, the origin and distribution of flared slopes are germane to the origin and evolution of bornhardts. The analysis of granite forms involves the consideration of some of the most dramatic and enduringly fascinating landscapes on Earth.

In this analysis, while personal views have not been suppressed, various, frequently opposed, interpretations have been outlined. No theory can ever be regarded as established, if only because the data on which it is based is incomplete or flawed. In our interpretations of various granite landforms we are enthusiastic and tenacious, but not dogmatic. While acknowledging the reality and wry humour behind the comment, we could not go so far as Lichtenberg when he remarked: *"instead of saying I have explained such and such a phenomenon, one might say, I have determined the causes for it the absurdity of which cannot be conclusively proved."*

Nevertheless, though the ideas presented appear soundly based in light of present data and concepts, our greatest reward will be to stimulate others with different experiences and backgrounds to formulate new explanations for the fascinating features found in granitic terrains. In this way, new light will be shed not only on the features discussed in this book, but on the evolution of landscapes in general.

Author Index

Ackermann E. 94, 106
Adams F.P. 33, 47
Agassiz L. 119, 150
Alexander F.E.S. 198, 204
Alexander L.T. 46, 47
Ambrose J.W. 79
Anderson A.L. 102, 106, 177, 204
Arsalan A.H. 266, 292

Bailey E.H. 171
Balk R. 16, 24
Barbeau J. 101, 103, 106
Barbier R. 122, 123, 147, 150
Barton D.C. 50, 62
Bateman P.C. 162, 167, 171
Beche H.T. de la 33
Bennett P.C. 251, 256
Berry L.R. 82, 107
Birkeland P.W. 62
Birot P. 8, 24, 110, 150
Blackwelder E. 72, 79
Blank H.R. 231, 232, 233, 266, 291
Blès J.L. 149, 150
Boissonnas J. 266, 291
Bornhardt W. 109, 110, 118, 120, 146, 150
Bourne J.A. 8, 24, 65, 72, 76, 79, 80, 98, 107, 110, 128, 151, 211, 234, 236, 253, 257, 266, 285, 292, 314, 325
Bowen N.L. 58, 62
Bowlby J.R. 292
Boyé M. 131, 132, 134, 150, 213, 233, 245, 256
Bradley W.C. 52, 62, 248, 256, 295, 305, 324
Brajnikov B. 42, 47, 125, 150
Branner J.C. 5, 8, 24, 32, 57, 62, 197, 204
Bremer H. 8, 24
Briggs H. 16, 24
Brook G.A. 149, 150
Brum A. de 156, 159, 172
Bryan K. 71, 79
Büdel J. 8, 24, 133
Bulow K. 193, 205

Caesalpus 10
Caillère S. 167, 171
Caldcleugh A. 255, 256
Campbell E.M. 8, 16, 24, 25, 29, 39, 47, 48, 128, 150, 159, 172, 204, 205, 222, 234, 266, 271, 292, 293, 324
Carlé W. 193, 205
Centeno J.D. 8, 24, 39, 48, 232, 233
Clarke E. de C. 178, 180, 205
Clayton R.W. 190, 205
Cloos H. 13, 24
Coates D.F 46, 47
Coates D.R. 281, 291
Cooper B.J. 305, 324
Corbin E.M. 207, 234
Cottard F. 91, 106
Cotton C.A. 193, 205
Coudé Gaussen G. 8, 24, 221, 233, 234
Cowie J.W. 78, 79
Cruden A.R. 266, 292
Chapman C.A. 33, 47
Choubert B. 8, 17, 24, 42, 45, 47
Choukrounne P. 22, 24

Dach W. 311, 324
Dale T.N. 14, 15, 24, 44, 45, 46, 47, 126, 150
Dardis G.F. 234
Davies J.L. 79, 256
Davis G.H. 16, 18, 24, 75
Davis S.N. 65, 79
Davis W.M. 64, 65, 79
De Prado C. 238, 256
Demek J. 155, 158, 171, 313, 324
Denham D. 46, 47
Dorn R.I. 251, 256
Drewry J. 246, 256
Dumanowski B. 190, 192, 205
Du Toit: see Toit
Dutton C.E. 120, 150

Engel C.G. 251, 256

Falconer J.D. 120, 130, 146, 150
Farmin R. 98, 106

Fell R. 34, 47
Finlayson B. 252, 256
Fisher G.W. 151
Fisher O. 120, 150
Folk R.L. 54, 58, 62
Frape S. 292
Friese F.W. 234
Fritsch P. 131, 132, 134, 150, 233, 245, 256
Fry E.J. 53, 62, 217, 234

Gagny C. 91, 106
Geikie A. 89, 106
Gelato G. 54, 62
Gèze B. 101, 103, 106
Gilbert G.K. 33, 34, 45, 47
Godard A. 8, 9, 24, 115, 150, 155, 157, 171
Goldich S.A. 60, 62
Grandal d'Anglade A. 255
Grawe O.R. 51, 62
Griggs D.T. 49, 62

Handley J.R.F. 155, 171
Hassenfratz J.-H. 86, 88, 106
Hedges J. 213, 234
Hellstrom B. 196, 205
Henin S. 167, 171
Herrmann L.A. 125, 140, 150
Heydemann M.T. 54, 62
Hibbard G.R. 145, 150
Hill S.M. 145, 150
Hills E.S. 16, 24, 314, 322, 324
Holmes A. 120, 126, 150, 171
Holzhausen G.R. 46, 47
Howard A.D. 69, 79
Huber N.K. 162, 171, 265
Hult R. 8, 24, 238, 256
Hume W.F. 169, 171
Hutton J.T. 52, 58, 62, 248, 256, 295, 300, 324

Iglesias M. 22, 24
Ikeda H. 9, 24, 142, 150, 246, 256, 280, 291
Isaacson E.de St.Q. 46, 47

Author Index

Jahn A. 155, 172
Jeje L.K. 141, 149, 150
Jennings J.N. 8, 24, 267, 291
Jessen O. 5, 109, 120, 137, 138, 150
Johnson R.J. 285, 291
Joly J. 53, 54, 62
Jones T.R. 218, 234
Joyce E.B. 145, 150
Judd W.R. 47, 48, 291
Jutson J.T. 119, 150, 185, 205, 248, 256

Kalk E. 251, 257
Kastning E.H. 111, 235, 256
Kendall P.F. 16, 24
Keyes C.R. 119, 150
Keyes D. 292
King L.C. 8, 20, 21, 24, 74, 79, 120, 137, 150, 207
Klaer W. 8, 24, 104, 106, 197, 205, 238, 243, 246, 248, 256
Kleman J. 119, 150
Kubicek P. 82, 107
Kvelberg I. 5, 24

Lagasquie J.J. 8, 24, 87, 96, 107, 115, 150
Lageat Y. 8, 24, 115, 150, 202, 205, 213, 234, 311, 313, 324, 325
Lamego A.R. 32, 42, 45, 47, 123, 129, 150
Larsen E.S. 59, 62, 102, 107
Le Conte J.N. 119, 127, 151
Leeman E.F. 99, 107
Leonard R.J. 287, 291
Lester J.G. 140, 151
Lichtenberg G.C. 330
Lindsay D.M. 300, 324
Lindsay D.S. 58, 62
Lindsay J.M. 76, 80
Linton D.L. 93, 107, 158, 172, 313, 324
Lister L.A. 96, 115, 140, 143, 151, 194
Livingstone D.A. 66, 79
Logan J.R. 196, 200, 201, 205
Loughnan F.C. 60, 62

Mabbutt J.A. 64, 79, 98, 107
Macaulay Lord 327
MacCulloch J. 58, 62, 83, 107
MacGregor P. 34, 47
Mackin J.H. 75, 79
Marcos A. 24
Martí C. 156, 172
Martín A. 8, 24

Matschinski M. 5, 24
Mayer R.E. 77, 79
McCarroll D. 53, 62
McClean R.J.C. 54, 62
McFall G.H. 266, 292
McFarlane M.J. 54, 62
McGee W.J. 69, 79
McGregor P. 34, 45, 47
McKay D.A. 292
McMartin I. 294, 325
McMillan R.K. 282, 292
Mennell F.P. 120, 127, 151
Merrill G.P. 33, 47, 130, 151
Meybeck M. 251, 257
Meyer B. 251, 257
Migin P. 82, 107
Migon P. 116, 126, 151, 311, 324
Mitra S. 123, 151
Mohajer A.A. 292
Monge C. 172
Monroe W.H. 303
Moon B.P. 234
Mountford C.P. 122, 151
Moyé D.G. 46, 48
Mueller J.E. 92, 159, 172, 177, 180, 185, 205, 298, 309, 324
Müller L. 46, 48
Mustoe G.E. 240, 257

Netoff D.I. 287, 291, 305, 324
Noble P.J. 54, 62

Oberlander T. 153, 172, 251, 256, 313, 324
Oen Ing Soen 119, 163
Ollier C.D. 8, 24, 34, 48, 60, 62, 84, 107, 145, 150
Owen L.A. 292

Pain C.F. 34, 48
Pallí Buxó Ll. 8, 25, 232, 234
Parga Peinador J.R. 22, 24
Parga Pondal I. 22, 24
Passarge S. 63, 79, 119, 120, 151
Patton E.B. 58, 62
Pearce M. 292
Pedersen K. 54, 62
Pedraza J. de. 8, 24
Peel R.F. 28, 75, 79, 88, 187, 189
Penck A. 238, 257
Pesci M. 79
Petit M. 8, 24, 196, 199, 205
Plug C. 266, 291
Plug I. 266, 291
Popoff B. 5, 24

Post van der L. 109, 151
Pugh J. 69, 79
Pye K. 202, 205, 213, 234

Rabenandrasana S. 180, 190, 196, 199, 205
Ramanohison H. 180, 190, 196, 199, 205
Read H.H. 10, 24
Reusch H.H. 5, 24, 238, 257
Richthofen F. von 22, 24
Rieper C. van 324
Ritchot G. 8, 24, 140, 144, 151, 155, 172
Robinson D.A. 148, 287, 292
Rodrigues L.M. 156, 159, 172
Rodríguez R. 23, 25, 139, 140, 151
Rognon P. 116, 146, 151, 157
Roqué C. 8, 24, 25, 232, 234
Russell R.S. 25, 172, 234, 292
Rutty A.L. 266, 292
Ruxton B.P. 72, 79, 82, 107, 193, 205

Sanz C. 8, 24, 25
Scott G.D. 217, 234
Scott W.B. 282, 292
Scrivenor J.B. 167, 172
Scheffer F. 251, 257
Schmidt-Thomé P. 193, 205
Scholz D.L. 197, 205
Schubert C. 271, 292
Schulke H. 287, 292
Schumm S.A. 151
Seager W.R. 161, 172
Sellier D. 202, 205, 213, 234
Serrat D. 156, 172
Shakespeare W. 145, 150
Shakesby R.A. 228, 234
Sharp R.P. 78, 79, 251, 256
Shroba R.R. 305, 324
Siever R. 58, 62
Simmons I.G. 155, 171, 172
Sjoberg R. 313, 325
Smith D.M. 65, 79
Smith L.L. 211, 234
Sosman R.B. 287, 292
Spencer-Jones D. 305, 325
Spooner G.M. 25, 172, 234, 292
Stapledon D. 34, 47
Stewart I. 292
St-Onge D.A. 294, 325
Streckeisen A.L. 9, 25
Sutton D.J. 282
Sved G. 280, 292
Swaine D.J. 87, 107

Author Index

Tarr R.S. 15, 25
Thomas M.F. 8, 25
Thomas R.L. 282, 292
Thomson J. 248, 249, 257
Toit A.L. du. 4, 21, 25, 78, 79, 109, 147, 151
Trendall A.F. 72, 79
Trudinger P.A. 87, 107
Tschang H.L. 197, 205, 244, 260, 292
Twidale C.R. 8, 9, 10, 16, 25, 29, 35, 39, 47, 48, 52, 58, 62, 65, 72, 74, 76, 79, 80, 82, 83, 88, 98, 107, 110, 123, 126, 128, 130, 137, 148, 150, 151, 157, 159, 172, 174, 177, 180, 183, 185, 202, 203, 204, 205, 207, 211, 213, 214, 222, 232, 234, 236, 248, 252, 253, 256, 257, 266, 267, 271, 280, 285, 291, 293, 298, 300, 311, 314, 324, 325
Twidale N. 72, 80, 314, 325

Ule W. 197, 205
Uña Álvarez E. de 8, 25

Van der Post see Post van der L.
Van Rieper III C.: see Rieper
Vegas R. 22, 24
Vidal Romaní J.R. 8, 9, 23, 25, 39, 48, 81, 82, 107, 123, 126, 151, 156, 159, 172, 180, 190, 196, 199, 205, 212, 232, 234, 236, 238, 246, 247, 252, 253, 257, 266, 285, 287, 292
Vilaplana J.M. 156, 172, 236, 257
Viles H. 53, 62
Vita Finza C. 292

Wahrhaftig C. 77, 78, 80, 162, 167, 171, 187, 294, 305
Wallach J.L. 266, 282, 292
Waters R.S. 155, 166, 172
Watson A. 202, 205, 213, 234
Watts W.W. 146, 151
Webb J. 252, 256

Whitaker C.R. 260, 292
White S.E. 52, 62
White W.A. 34, 48
Whitlow J.R. 115, 151, 228, 234
Wilhelmy H. 9, 25, 193, 202, 205, 249, 257
Williams G.E. 75, 80
Williams M.A.J. 79, 256
Williams R.B.G. 148, 287, 292
Willis B. 109, 120, 151
Winkler E.M. 49, 62
Worotnicki G. 46, 47
Worth R.H. 8, 25, 155, 172, 211, 219, 234, 264, 292
Wray D.A. 120, 150

Yepes, J. 23, 25
Young A. 294, 325
Young R.W. 294, 325

Zernitz E.M. 18, 25
Zêzere J. 156, 159, 172

Location Index

A Coruña, Galicia 97, 99, 123, 245, 316, *318*
Africa 75, 193,194, 251, 266, 289, 296, 313
Adamello Alps 11
Air Mountains, southern Sahara 125
Albany, Western Australia 157, *186*
Alice Springs, central Australia 153, 155, 302, *305*, 313
Aliena, The Flinders Ranges, South Australia *296*
Altar de Cabrões, Serra do Gerêz, northern Portugal *231*
Amboira, Angola *138*
Amboromena Pic, Andringitra Massif, Madagascar *221*
American Southwest 69, 251, *296*, 297
American West 69, 120, *296*
Ampidianambilahy, Andringitra Massif, Madagascar *70*, *175*, 190, *225*
Andén Verde, Canary Islands 306, *311*
Andorra 156
Andorran Pyrenees *156*
Andringitra Massif, Madagascar *8*, *70*, *175*, 180, 190, 196, *199*, 221, *225*, 236
Angola *5*, 120, 137, 138, 149, 196
Anillaco, República Argentina 255
Antarctica 243, *244*, 248, *297*, 310
Antilles 56, 60
Appalachians, USA *297*, 298
Arabia 126, *296*
Arizona 69, *92*, 251, 305
Armchair, The Flinders Ranges, South Australia 126
Arnhem Land, Northern Territory 294
Arran, Scotland 297

Arteixo, A Coruña, Spain 316, *318*
Ashburton River, Western Australia 225
Aswan, Egypt 11
Atlantic coast of Galicia 34, 314
Atlantic Ocean 23
Augrabies, Northern Cape, South Africa 95, 98, 100, 103, 105, 106, 280, 285
Australia *4*, 64, 76, *83*, 93, 137, 146, 149, 190, 192,193, *197*, 204, 211, 219, 221, 238, 243, 248, 251, 266, *277*, 289, 295, 296, 297, 310, 313
Australian deserts 260
Ávila-Villacastín, central Spain 72, 73, 75
Ayers Rock, (Uluru), Central Australia 122, 126, 180, *294*, 295, 296, 297, 298, *300*, *302*, 303, *308*

Baia de Guanabará, Rio de Janeiro, Brazil 119
Balanced Rock, Llano, central Texas 93
Balancing Rock, Harare, Zimbabwe 93, *94*
Balladonia, Western Australia 297
Bangalore, Peninsular India 131, 133
Barbanza Uplands 235
Barna Hill, South Australia 298
Baxter Hills, Eyre Peninsula, South Australia 298, 308
Beardown Man *166*, 166
Beda valley, South Australia 297
Benbarber Corner, northwestern Eyre Peninsula, South Australia 176
Black Forest, Germany 255
Black Hill, Western Murray Basin, South Australia 307

Blackingstone Rock, eastern Dartmoor, England 14, 42, *44*, *133*, 171
Bloedkoppie Dome, central Namibia 245
Bohemia 158, 196
Bohemian Massif 155, 219, 245, 313
Bongoberg, Angola *138*
Boone's Cave, North Carolina, USA 237
Boulder Rock, Darling Ranges, Western Australia *290*, 291
Bowerman's Nose 95
Brazil *5*, *30*, 39, 42, 60, 65, 109, 119, 140, 196, 197, 211, 255, *296*
Britain 83, 211, 249
Brittany *41*, 202, *203*, 213, 255, 313
British Columbia, Canada *323*
Broken Hill, New South Wales 298
Bruce Rock, Western Australia *226*
Buccleuch, south of Johannesburg, South Africa 131, *214*, 284
Bungle Bungle Ranges, Western Australia 294, *296*
Bushman's Kop *118*
Bushmanland Surface, Western Cape Province 74, *296*

Cairns-Mosman area, north Queensland 19
Calca Hill, Eyre Peninsula, South Australia 131, 180
Calca Quarry 131, 180
California 68, 102, 237
Caloote, Murray Basin, South Australia *187*, *226*
Camelle, Galicia, northwestern Spain 103
Cameroon, West Africa 77, 214

336 Location Index

Canada 2, *15*, *45*, 73, 83, 163, *165*, 282, *296*, 313, 322, *323*
Canadian Arctic 79, 219
Cape Fold Belt, South Africa *296*
Cape of Good Hope, South Africa 78, 297
Cape Paterson, eastern Victoria *298*
Cape Town, South Africa 60, *61*, 78, *104*, *141*, 196, *197*, *208*
Cape Vilano, Galicia, northwestern Spain 103
Cape Willoughby, Kangaroo Island, South Australia 103, 314, 316, *318*
Cape Wilson, South Victoria 315
Carappee Hill, northeastern Eyre Peninsula, South Australia 267, *269*
Carpentaria Plains, Queensland *296*
Cash Hill, Eyre Peninsula, South Australia 270, *272*
Cassia City of Rocks 177, *180*, *185*, *202*, 288
Castle Rock, Mt Manypeaks, Western Australia 156, *157*
Catalonia, Spain 232, *297*, *298*, 310
Cathedral Rocks, Yosemite, Sierra Nevada, California 162, *164*, 313
Cávado, Portugal 22
Cavallers Dam, northeastern Pyrenees *3*
Central Massif of France 72
Cerro Prieto, Toledo, Spain 294
Cerro Torre, Patagonia, Argentina 313
Cerro de Moraján, Toledo, Spain 294, *300*
Charnwood, English Midlands 146
Chazeirollettes, southern Massif Central, France 88, *263*
Childara Rocks, eastern Victoria Desert, South Australia *214*
Chillagoe, north Queensland *296*
Chilpuddie Hill, Eyre Peninsula, South Australia 180, *184*, 193
China, Kweilin *296*
Cíes Islands, western Galicia, Spain *109*
City of Rocks, New México *296*, *297*, 305, *309*
Ciudad Encantada, Cuenca, central Spain *223*, 223, 298

Coast Mountains, British Columbia *21*
Coles Bay, Tasmania 322
Colorado Plateau, Utah *295*, *298*, 305, *310*
Colorado, USA 53, 75, 87, 140, 142, 235, 287, *296*, *297*, 298, 305
Córdoba, República Argentina 255
Costero-Catalana Range, Spain 245
Coolamon Plain, New South Wales 303
Corrobinnie Depression, South Australia 190
Corrobinnie Platform, Eyre Peninsula, South Australia *191*, 224, 287, 289
Corrobinnie Hill, northwestern Eyre Peninsula 69, 190, *283*, 285, 287, 289, *296*
Corsica 8, 58, 103, 238, 243, *245*, 246, 248, 310
Costa Brava, Catalonia 8, *241*, 246
Costa da Morte, Galicia, northwest Spain 103, 155, *323*
Costa Rica 119
Crowder Rocks, Minnipa-Wudinna district, Eyre Peninsula, South Australia 227
Cue, Western Australia 74
Curtinye Hill, Eyre Peninsula, South Australia 298
Christopherus-Stein, Blockheide Gmünd-Eibenstein, Gmünd, lower Austria 265

Daadenning Hill, Merredin, Western Australia *283*
Dana Plateau 167
Darling Ranges, Western Australia 131, *290*, 291
Dart River 166
Dartmoor, southwestern England 7, 8, 18, 19, 32, 36, 41, 42, 72, 75, 94, 95, 99, 185, 217, 218, 219, 240, 245, 264, 265, 266, 289
Darwin, Northern Territory 55, 255, *297*, *298*, 308
Death Valley, California 298
Devil's Marbles, Northern Territory 50, 83, *84*, *94*, *112*, *158*, 159, *160*, 171, 172, *184*, *197*, *261*, *297*, 313

Domboshawa, Zimbabwe 56, 56, 196, *220*, 221, 228, *229*, *291*, 291, 327, 328
Domeland, central Sierra Nevada, California *117*, 140
Douro, Portugal 22
Drakensberg, South Africa *296*, *297*, 303, 305, 306, *308,* 311, 312
Dumonte Rocks, Minnipa-Wudinna district, Eyre Peninsula, South Australia 227, *228*

East Africa 110
Eastern Cape Province, South Africa *311*, *312*
Ebaka dome, South Cameroon 131, *134*, 213, 214, 245
Egypt 11, 49, 169, 190, *192*, *297*
Egyptian Desert 190, 192, 289
Elkington Rock, Eyre Peninsula, South Australia 131, *134*, 183
Enchanted Rock, central Texas *104*, *111*, *230*, 236, *236*, *237*
Encounter Bay, South Australia 315
England *19*, 32, 36, *44*, 46, 296, 313
English Pennines – see Pennines, England *296*, *297*, 302, 303
Esperance, Western Australia 314
Europe 8, 66, 296
Everard Range, South Australia 113, *113*, 145
Eyre Peninsula, South Australia 16, 49, *53*, *56*, 65, 69, 72, 174, 180, 183, 196, 203, 214, 227, 230, 232, 245, 251, 260, 266, 270, 271, 282, 296, 297, 298, 308, 310, 314, 317, 322
Ézaro, Río Xallas, A Coruña, northwestern Spain *241*, 245

Fafião Valley, southern Galicia *59*
Falkland/Malvinas Islands *298*
Fennoscandia 147, 149
Finland 42, 116
Fitzroy Basin, Western Australia *296*
Flinders Ranges (The), South Australia 126, 169, *170*, *296*, *297*, *298*, *306*, 306, 307–8
Fontainebleau, France 298
France *41*, 72, *88*, 88, *92*, 103, 119, 155, 202, *203*, *263*, *298*, 313

Location Index

Freeman Hill, Eyre Peninsula, South Australia 60, *61*
Free State, South Africa 20–1, 131, 134
French Guyana *17*, 17, 42, *45*, 58, 125
French Pyrenees *81*
Frog Island, (Pulau S'kodo or Sekudu), Singapore *200*

Galicia, Spain 23, *27*, *28*, 39, *43*, 51, *59*, 93, 94, 97, 99, *100*, 103, 221, 232, 238, *238*, 289, 297, 298, 311, 322, *323*
Galician Massif 8
Galong, New South Wales *297*, 307
Garies, Namaqualand, (Western Cape Province), South Africa *105*, 131, *133*
Gawler Ranges, South Australia 8, 16, 29, 70, 71, 295, 296, 297, 298, 299, 301, 302, 303, 304, 305, 306
Gebel Harhagit, Egypt 190, *192*
Georgia, USA 45, 125, *125*, 266
Gilbert Dome, Sierra Nevada, California 119
Girona, Costa Brava, northeastern Spain 72, 73, 211, *241*, 246
Girraween National Park, New South Wales 235
Glasshouse Mountains, Queensland 298
Glennie Group, Victoria 317
Gmünd, Lower Austria 265
Grampian Ranges, western Victoria 293
Gran Sabana, Brazil *296*
Grand Canyon, Western USA 78
Granite Mountains, Southern California *68*
Great Plains, Montana 65
Greenland *78*, 79, 119, 162, *163*
Groot Spitzkoppe, central Namibia *110*
Guadarrama, Madrid, Spain 238
Guitiriz, northwestern Galicia, Spain 39, 40, 44, 125, 288
Guyana 235, 266, *296*

Halls Creek, northwestern Australia 260
Hamersley Surface 155
Hanging Rock, Victoria 303
Harare, Zimbabwe 56, *85*, *94*, 131, 220, 228, *229*, *230*, 291, 327, 328

Haytor, Dartmoor, southwestern England 58, *59*, 155, *156*
Heltor, Dartmoor, southwestern England *30*
Herbert River Falls, north Queensland *84*
Hercynian Uplands, Western Europe 159
Hesperian Massif, Iberia 83
High Tatra Mountains, Poland 236
Hillock Point, Yorke Peninsula, South Australia 315
Hoggar Mountains, central Sahara 255
Hombori Massif 294
Houlderoo Rocks, Gawler Ranges, South Australia 70, 71, *106*, 159
Hong Kong 243, *244*, *296*, 313
Hyden Rock, Western Australia 137, *173*, *174*, *175*, 211, *263*, 285, *286*

Iberian Peninsula 8, *22*, 22, 72, 75, 158, 159, 196, 219, 227, 233, 260, 296
India 4, 49, 131, 149
Indian Peninsula 9
Indonesia 60, *296*
Investigator Group, Great Australian Bay *16*
Ipoh, West Malaysia *303*
Iran 120
Italy 11, 34, *297*

Japan 69
Johore Strait 196, 200
Joshua Tree National Monument, California 35, *36*

Kamiesberge, Namaqualand (Western Cape Province) 113, *114*, 140
Kap Ingersoll, North Greenland 78
Karibib, central Namibia 60, *60*, *153*
Karimun Island, western Indonesian Archipelago *85*
Karkonosze Mountains, Poland 58, 155, 236, 311
Karroo 21
Katajuta – see The Olgas 126, 295, *296*, *299*, *301*, *303*, 308
Kazan Region, Canada *2*
Keetmanshoop, South Namibia *296*

Keith, South East district, South Australia *183*
Kennedy Rock, Matopos Hills, Zimbabwe *96*
Kenya 140
Kestor, Dartmoor, southwestern England *218*
Kimba, Eyre Peninsula, South Australia *296*, *298*, 298
Kimberleys, Western Australia 294
King Rocks, Hyden, Western Australia 285, *286*
Kings Canyon, Northern Territory *296*
Kobe/Mt Rokka, Japan *281*
Kogon Dutsi, Nigeria 130
Kokatha, Gawler Ranges, South Australia 264
Kokerbin Hill, Western Australia 137, *239*, 245, *252*, *267*, *277*, *297*
Kolar Goldfield, southern India 46
Kondinin Rock, Western Australia *117*
Korea 69, 246
Kosciusko Mountains, soutwestern New South Wales 260, *261*
Kråkmotinden, northern Norway 116
Kulgera Hills, Northern Territory 208, 266, 273, *277*, 280, *298*
Kwaterski Rocks, Eyre Peninsula, South Australia *178*, *209*, *215*, *218*, 232, *297*
KwaZulu/Natal *118*, 225
Kweilin, China *296*
Kylie Lake, southern Zimbabwe *20*

Labertouche Cave, Neerim South, Victoria 235
Labrador High Plain, eastern Canada 73
Labrador Plateau, Canada 163, *165*
Labrador Trough 163
Labrador, Canada *15*, 73, 120, 163, 266
Labrador Peninsula 73
La Clarté, Brittany, France *41*
La Pedriza, central Spain 8
Lake Greenly, southern Eyre Peninsula, South Australia 185
Lake Tchad, central Africa 99, *101*
Laxe, Coruña, Spain 217

Leeukop, Free State, South Africa 131, *134*
Lézaro, Galicia, Spain *23*
Lidbergsgrottorna, northern Sweden 314
Lightburn Rocks, eastern Great Victoria Desert, South Australia 216, 270
Lima, Galicia, Spain 22
Linares, Andalusia, Spain 123, 146, *148*, 211
Litchfield National Park, Darwin, Northern Territory 255, 298, 308
Little Shuteye Pass, Sierra Nevada, California 37
Little Wudinna Hill, Eyre Peninsula, South Australia *105*, 142, *227*, 271, *275*, 276, 280
Llano, central Texas *93*, *104*, 153, 230, *230*, 236
Lleida Province, Spanish Pyrenees 236
Lost Creek Granite Caves, Colorado 235
Louro Uplands, northwestern Spain *212*, 235, *253*
Louro, A Magdalena, Muros southern Galicia, Spain 217
Lugo, Galicia, northwestern Spain *14*, *86*, *124*
Lundi River Bridge, southern Zimbabwe *194*

Mac Donald, northwestern Canada 2
Madagascar *8*, 60, *70*, *175*, 180, 190, 196, *199*, *221*, *225*, 236, 255
Malay Peninsula 200
Malaysia 196, *201*, *296*, *297*
Mali 294, *296*
Malmesbury, Western Cape Province, South Africa *141*
Malvinas Islands *298*
Marehuru Hills, southern Zimbabwe *111*
Margeride, central France *167*, 167, 196
Mariz Quarry, Galicia, northwestern Spain *31*, 36, *39*, *40*, 280, *288*, 288, *298*
Massif Central of France 72, 86, *88*, 155, 167, *263*, 296, 313
Matopos Hills, Zimbabwe *96*
Mauritania *28*
Mallos de Riglos, Pyrenees, Spain 295

Mdzimba Hills, Swaziland 213
Meckering, Western Australia *1*
Meda de Rocalva, Serra do Gerês, northern Portugal *161*
Mediterranean 62, 69, 149, 310
Meekatharra, Western Australia 74, *75*, *296*
Merced River valley, Sierra Nevada, California 167
Meseta, Iberian Peninsula 72
Meteora, Greece 295, *296*, *301*
Mexico *161*, 162, *163*, *296*, *297*, 298, 305, 309, 313
Midrand, South Africa 131, *135*, 213
Mingan-Natashquan, Québec, Canada *45*
Minnipa Hill, northwestern Eyre Peninsula *31*, *34*, *57*, 131, *135*, *178*, *182*, 216, 232, *250*, 259, *277*, *279*, 281
Miño, Galicia-northern Portugal 22
Mistor Pan, Dartmoor, southwestern England 219
Mitchell's Nob, Musgrave Ranges, South Australia 126
Mojave Desert, southern California 153, 190, 313
Monolith Valley, New South Wales 294
Monte Louro, Galicia, northwestern Spain *212*, *237*, *253*
Montserrat, northeastern Spain 295
Monte Pindo, Galicia, Spain 103
Monzoni, Tyrol 11
Mozambique 120, 126, 171
Mt Bundey, Darwin, Northern Territory 55
Mt Conner, central Australia 126
Mt Hall, Eyre Peninsula, South Australia 232, *240*
Mt Kobe, Mozambique 171
Mt Kosciusko, New South Wales 260, *261*
Mt Lindsay, Mann Range, central Australia *110*
Mt Lofty Ranges, South Australia *6*, *51*, *89*, *95*, 156, 260, *297*
Mt Magnet, Western Australia *260*, 270, *272*, *280*, 281, 285, 287, *288*, *298*
Mt Manypeaks, Albany, Western Australia *157*, 157, *186*
Mt Monster, South Australia 93
Mt Rokka, Japan 281

Mt Sinai, Sinai Peninsula, northeastern Egypt 129
Mt Ward, Northern Flinder Ranges, South Australia 126
Mt Whitney, East Sierra Nevada, California *296*, 313
Mt Whitney, Southern Greenland 162
Mt Wudinna, Eyre Peninsula, South Australia 30, 35, *38*, 282
Murphys Haystacks, Eyre Peninsula, South Australia *56*, *128*, *186*, *196*, 245, *296*
Murray Basin, South Australia *187*, *226*, *307*
Murray Valley, South Australia 76
Musgrave Ranges, South Australia 126, *127*

Namaqualand (Western Cape Province) 7, *67*, 68, 76, *81*, *82*, *105*, *112*, *114*, 131, *133*, 146, *147*, *154*, *189*, *296*, 313, *328*
Namibia *3*, *6*, 65, 66, 67, 68, 125, 126, 149, 248, 280, *296*, 310
Namibia-Namaqualand border district *147*
Naraku, northwest Queensland *188*
Neue Smitsdorp, Northern Transvaal 131
Nevada, Utah, Colorado *296*
New England 46
New South Wales 46, 51, *52*, *85*, *90*, *91*, 99, 219, 235, 255, 260, *261*, 263, 280, *286*, 289, 294, 298, 303, *304*, *307*
Newfoundland 149
Ngoura and Gamsous, Lake Tchad region, central Africa *101*
Nigeria 69, 120, *141*, 149
Nile Valley 50
North Carolina, USA 42, 237
Northeast Sardinia 243
Northern Province, South Africa 65, *75*, *296*
Norway 116, 119
Nowra, New South Wales 294
Nullarbor Plain, Western Australia and South Australia *296*
Nuwerus, Namaqualand (Western Cape Province) 131

O Cadramón, Abadín, Lugo, northwestern Spain 72

Location Index

O Pindo Galicia, northwestern Spain 238
Okement River, Dartmoor, southwestern England 166
Olgas – see The Olgas 126, 295, 296, 299, 301, 303, 308
Ontario, Canada 282
Organ Mountains, New Mexico 161, 162, 163, 296
Oyo Region, Nigeria 141

Paarl Mountain, Western Cape Province, South Africa 196, 297
Paarl, Western Cape Province, South Africa 297
Paarlberg Quarry, Western Cape Province, South Africa 60, 61, 104, 197, 208
Palmer, eastern Mt Lofty Ranges, South Australia 6, 51, 89
Palmerston, South Island, New Zealand 298
Panticosa, Spanish Pyrenees 52, 82, 93
Panticosa and Cavallers, Spanish Pyrenees 93
Pão de Açucar, Rio de Janeiro, southeastern Brazil 30, 58, 109, 116, 122, 123
Papua New Guinea 33, 169
Parthenon, Greece 248
Parys, South Africa 21
Patagonian Andes 162
Payne's Find, Western Australia 276
Pearson Islands, Investigator Group, Great Australian Bight 16, 180, 313
Peella Rock, South Australia 69, 190
Peninsular India 131, 149
Pertnjara Hills, Northern Territory 302, 305
Peruvian Andes 169
Peyro Clabado, Sidobre, Southern France 92, 243
Pic Boby, Andringitra Massif, Madagascar 8, 180, 199
Pic Parana, southeastern Brazil 122, 122
Pietersburg, Northern Province 131, 245
Piedmont, USA 120
Pilbara region, Western Australia 70, 98, 146, 153, 153, 154, 155, 260, 266, 296
Pildappa Rock, Eyre Peninsula, South Australia 18, 117, 131, 174, 194, 203, 208, 209, 210, 274, 278
Pindo, northwestern Spain 28, 35, 43, 235, 238, 245
Pine Creek, Northern Territory 87
Pitões das Junhas, Serra do Gêrez, northern Portugal 155, 162, 162, 296, 313
Plateau de Andohariana, Andringitra Massif, Madagascar 196
Platja d'Aro, Girona, Spain 72, 73
Podinna, Eyre Peninsula, South Australia 297
Point Brown, Eyre Peninsula, South Australia 297, 315, 316, 319
Point Drummond, South Australia 322, 324
Poland 58, 60, 82, 211, 236
Polda Rock, Eyre Peninsula, South Australia 142
Pomona Quarry, Harare, Zimbabwe 131
Poondana Rock, Minnipa, Eyre Peninsula, South Australia 138
Porriño Quarry, Galicia, Spain 13, 27
Port Hedland, Western Australia 147
Portela d'Home, Portugal-Galicia 50
Port Kenny 176
Portugal 93, 221, 232, 255, 298
Pretoria, South Africa 74, 75, 131, 135, 214
Puerto Rico 303
Pulau S'kodo – see Frog Island 200
Pulau Sekudu – see Frog Island 200
Pulau Ubin, Johore Strait 196, 197, 200, 200, 201
Pyrenees 3, 8, 52, 81, 82, 93, 156, 156, 159, 235, 236, 260, 262, 263, 295, 296, 313

Quarry Hill, Eyre Peninsula, South Australia 35, 38, 131, 177, 180, 273, 275, 278
Quarry, Barre, Vermont, USA. 39, 43
Quebec, Canada 45
Queensland 4, 6, 19, 235, 298
Quenast, Belgium 46

Red Sea Hills, Egypt 169
Reedy Creek, Western Murray Plains, South Australia 76
Remarkable Rock, Kangaroo Island, South Australia 99, 100, 158, 195, 242, 297, 315
Reynolds Range, central Australia 112
Rio de Janeiro 5, 30, 39, 45, 58, 109, 116, 119, 122, 123, 129, 140, 169, 169, 255, 314
Río Tabalón, Guadarrama, Central Spain 41
Río Vilamés, Serra do Xurés, Southern Galicia, northwestern Spain 231
Río Xallas, Galicia, Spain 23
Robe River 155
Rock of Ages Quarry, Barre, Vermont 39, 43
Rocky Mountains, Boulder, Colorado 53, 75, 87, 140, 142
Rooiberg, Namaqualand (Western Cape Province), South Africa 76, 112
Rooifontein Valley, central Namaqualand (Western Cape Province) 76
Roopena, Eyre Peninsula, South Australia 296
Roper River, Northern Territory 294
Roques Planes, Girona, Spain 179
Roraima Plateau, Venezuela 54, 255, 303
Roughtor, Bodmin Moor, SW England 29
Rupununi savannas, Guyana 235

Sabah, East Malaysia 162, 164, 168, 266
Sahara Desert 28, 69, 116, 187, 188, 297, 298
Saldanha, Cape Province, South Africa 64, 194
Salinaland division, Western Australia 185
San Andreas Fault, California 282
Sarawak, Malaysia 297
Sardinia 88, 243
Scandinavia 128, 196, 219
Scrubby Peak, Gawler Ranges, South Australia 306
Scholz Rock, Eyre Peninsula, South Australia 245
Scotland 146, 148, 297

Location Index

Seoul, South Korea 237
Serra da Estrela, central Portugal 153
Serra do Gerês, northern Portugal *103*, 159, *162*, *165*, 221, *231*, 233, 282
Serra do Xurés, Galicia, Spain 99, *100*, *231*
Seychelles 9
Shag Point, Moeraki, South Island, New Zealand 317, *321*
Ship Rock, New México 298
Siberia 65
Sierra de Gredos, central Spain 155
Sierra de Guadarrama, central Spain *186*, 232, *233*, 238, *298*
Sierra Nevada, California 4, 10, 11, *29, 31, 36, 37*, 59, 75, *77, 77, 117*, 119, 127, 140, 142, 162, *164*, 167, 180, *187*, 217, 297, 313
Sinai Peninsula 129, 169
Singapore *164*, *168*, 196, *200*
Sisarga Grande, Malpica, Coruña, northwestern Spain *115*
Slippery Hills, northern Namibia 109
Smooth Pool, Eyre Peninsula, South Australia 314, *316*, 322
Snowy Mountains, New South Wales 46, *51*, *52*, 83, *85*, *90*, *91*, 99, 219, 263, *286*, 289
Sonoran district 69
South East district, South Australia *12*, 93, *183*, 221, *222*, 245
South East Asia 153
Southern Ontario 46
Spain 211, 213, 255, 297, 298
Springbok Flats, Pretoria 74, 75
St John's Peak, Sabah, East Malaysia *164*
St Uzek, Britanny, France 202, 203, 213
Stone Mountain, Georgia 45, *125, 125*
Sudan 190
Sudeten Mountains, southern Poland 126
Sugarloaf, Organ Mountains, southern New México *161*
Sullivan Rock, Darling Ranges, Western Australia 131
Surinam 196, 211
Swakop valley, central Namibia *154*

Swakoprivier, central Namibia 153
Swaziland 196, 202, 213
Sydney Basin, New South Wales 255, 294
Syene 11

Taj Mahal 248
Talia, Eyre Peninsula, South Australia *297*, 298, 316, 318, *319*, *321*, *322*
Tamar River, Dartmoor 166
Tambre 22
Tampin, West Malaysia 55, *86*, *169*, *195*, *201*, 260, *262*
Tanganyika 155
Tanzania 146
Tassili Mountains, Algeria 146, *147*
Tatra Mountains, Poland 236
Tavy River, Dartmoor 166
Tcharkuldu Hill, Eyre Peninsula, South Australia 232, *250*, *252*, *254*, *284*, 285, 287, 289
Teign River, Dartmoor 166
Tenaya Lake, Yosemite, Sierra Nevada, California 35, *36*
Tent Hills, South Australia 189
Texas Canyon, Arizona *92*, 266, 297
The Granites, Mount Magnet, Western Australia *243*, 260, 270, 272, 285, 287, 288, 289
The Humps, southern Yilgarn, Western Australia *145*, 190, *209*
The Leviathan, Buffalo Mountains, Victoria *83*, 83
The Needles, South Dakota *296*, *312*, 313
The Olgas (Katajuta), Northern Territory 126, 295, *296*, 299, *301*, 303, 308
The Pinnacles, New South Wales 298, *304*
Thompson River, Rocky Mountains, Colorado *142*
Tibooburra, New South Wales *297*
Tindal Plain, Northern Territory 254
Tocumweal, New South Wales 280
Tolmer Rocks, South East district, South Australia 221, *222*
Tonale Alps, northern Italy 11
Tooma, Snowy Mountains, New South Wales *91*, 99

Touça River, Serra do Gêrez, northern Portugal 103
Traba Massif, Galicia, northwestern Spain 139, *140*, *202*, 254
Tsodilo, northern Namibia 109
Tungkillo, Mt Lofty Ranges, South Australia *95*
Turtle Rock, Eyre Peninsula, South Australia *181*, 203, *204*, 251

Úbeda-Linares, Andalusia, southern Spain 123, 146, *148*
Ucontitchie Hill, Eyre Peninsula, South Australia 32, 39, 40, *42*, *65*, 66, 72, 131, 137, 174, 176, 199, *220*, 239, 243, 251, 279
Uluru – see Ayers Rock 122, 126, 180, *294*, 295, *296*, *297*, *298*, *300*, 302, 303, 306, *308*, 308
Ulla Galicia, northwest Spain 22
Umgeni River, KwaZulu/Natal, South Africa *118*, 140, *225*
Umia, Galicia, northwest Spain 22
Umtata, Eastern Cape Province, South Africa *306*, *308*
Usakos, central Namibia 66, 67
Utah, USA 296, *298*, 305

Vaal River, South Africa 20–21, *21*
Valley of Thousand Hills, KwaZulu/Natal, South Africa 140
Vanrhynsdorp, Namaqualand (Western Cape Province) 131
Varley Township Hill, Western Australia 190, *191*
Veyrières, southern France *178*, 180
Victoria *83*, 196, 235, 240, 303, 315, *317*, *322*
Vigo Ría, Galicia, northwestern Spain *109*
Villacastín, central Spain 72, *73*, 75
Volzberg, Surinam 196
Vredefort, Free State 20, 21, 131
Vredenburg, Western Cape Province, South Africa 197

Wallala Hill, South Australia 287, 289
Wattern Tor, Dartmoor, southwestern England *166*

Location Index

Wattle Grove Rock, Eyre Peninsula, South Australia *297*
Waulkinna, Gawler Ranges, South Australia 70
Wase Rock, Nigeria 298, *304*
Wave Rock, Hyden Rock, Western Australia *173*, 174, *175*, 203
West Africa 77, 119, 190, 214, *296*
West Malaysia *55*, *86*, *169*, *195*, *201*, 260, *262*, *303*
West Beach, Esperance, Western Australia 314
Western Australia *64*, 65, *70*, 74, 98, 175, 190, 296, 297, 298, 314
Western Cape Province 65, 197
Western Cordillera, USA 235

Whites Quarry, Minnipa, Eyre Peninsula, South Australia 31, *34*
Wilcannia, New South Wales *296*
Wilsons Promontory, Victoria 145, *317*
Windmill Bay, Kangaroo Island, South Australia 314, *315*
Witteklip, Western Cape 197
Wombeyan Caves, New South Wales 303
Wudinna Hill, Eyre Peninsula, South Australia 30, 35, 53, *117*, 129, 131, 137, 196, 197, *198*, *227*, 251, 267, 269, *270*, *271*, 273, 274, 275, 276, 278, 280, 281, 282, 285
Wyoming, USA 75, *298*, 303

Xallas Río, Galicia, Spain see Río Xallas 22, *23*, 23, *241*, 245

Yarwondutta Reservoir *136*
Yarwondutta Rock *57*, 133, 135, *136*, 136, *177*, 180, *182*, 192, 200, 203, 204, *212*, *216*, 216, 251, *297*
Yilgarn Block, Western Australia 145, 238, 266, *286*, 296, *298*
Yilgarn Craton, Western Australia 119, 239, 296, 298
Yilgarn Province 1, 180
Yosemite, Sierra Nevada, California 11, *29*, *31*, 35, *36*, 127, 140, 162, *164*, 217, 313
Yukon, Canada *296*

Zebra Mountains, central Namibia *242*
Zimbabwe 42, 45, 179, 230, 297, 298, 313, 327, 328

Subject Index

acanaladuras 224
acicular towers 163, 164
acquired permeability 15
acuminate, or pointed, form 162, 186
adamellite 11
alcove 59, 238, *240*, 249
algal coatings *183*, 184
all slope topography 168, 169
all slope topography – development *171*
allotriomorphic 11
alveolar weathering 240, *242*
alveole 53, 238, 240, *242*, 243, 248, *297*, *322*
angular drainage pattern 18
anhedral 11
anomalous drainage
anomalous drainage pattern 20
anomalous drainage due to – antecedence 20
anomalous drainage due to – diversion 20
anomalous drainage due to – inheritance 20
anomalous drainage due to – stream persistence and valley impression 20
anomalous drainage due to – superimposition 20
antecedence 20, 22
antiform 126, 129, 142, *144*, 155, 327
anvil rock 186, *187*
aperture 54, 235, *238*, 238, 250
aplite 11, 13, 61, 197
araceenhorst 211
arène 83
armchair shapped hollow: see rock basin
A-tents
A-tents – contemporary 280, *327*
A-tents – crestal orientation *271*, 280, 282
A-tents – definition 266
A-tents – description 266–71
A-tents – elongate 267, *270*, 282

A-tents – origin 278
A-tents – pressure ridges 282
Augen *12*
azonality 293–308, 309–13

bacteria – case-hardening 250–1
balanced rock *93*, *94*, 222
basal freeting 181–8, 190, 204, 245, 302
basal steepened slope 132
basin: see rock basin
bathtub 211, *214*
batholit 10, 16, 22, 23
batholit – fractures *13*
Baumverfallspingen 211
Bauxite 49
Bergfussniederungen 190
Benitiers 232
billiard-table surfaces 185
books 243
blade-like or ensiform, residual 186
bornhardt 109, 110, 116, 129, 167,173
bornhardt-bevelled 115–18, *115*, 122, *300*
bornhardt – characteristics 110, 115–18, 140–2, 317
bornhardt – conglomeratic *301*
bornhardt – dacitic 128
bornhardt – description 110
bornhardt – quartzitic 298, 303
bornhardt – sandstone 122, *296*
bornhardt – theories of origin 32–47, 118–31
bornhardt – theories of origin – environmental 32–47, 118–30
bornhardt – theories of origin – scarp retreat 120–21, *121*, 131, 140
bornhardt – evidence and argument 120–49
bornhardt – evidence and argument – age of residual 142–5
bornhardt – evidence and argument – antiquity 149

bornhardt – evidence and argument – coexistence of forms 140
bornhardt – evidence and argument – contrasted hill and plain 131
bornhardt – evidence and argument – contrasted fracture densities 127, 132
bornhardt – evidence and argument – exhumed forms 146
bornhardt – evidence and argument – flared forms 132, 136
bornhardt – evidence and argument – fracture defined margin 121, 142
bornhardt – evidence and argument – incipient domes *118*, 131
bornhardt – evidence and argument – minor forms – subsurface 131–2
bornhardt – evidence and argument – multicyclic landscapes 140–2
bornhardt – evidence and argument – pattern in plan 117, 139–40
bornhardt – evidence and argument – stepped inselberg 132–38
bornhardt – evidence and argument – topographic setting 140
borrageiros – see castle koppie 165
boulder 83, 84, 175, 263
boulder – and epigene origin 103–4
boulder – boule 83
boulder – distribution 83, 103–4
boulder – etching 81–2
boulder – floater 83
boulder – fluted 173, 193, *195*, *196*, *200*
boulder – heart of the block 83
boulder – kernel 83

Subject Index

boulder – morphology 83
boulder – names 83, 211, 224
boulder – norite 58, 59, 300, 302, *307*
boulder – occurrence 83
boulder – orthogonal fractures 83–95
boulder – rounding 86
boulder – shape 83, 90–99
boulder – size 90–99
boulder – subsurface development 83, 88, 90, 91, 93, *104*, 105, 106
boulder – two-stage mechanism 81–83
boulder beach 103
boulder field *168*
bushfire 279, 280, 286, 289
Bushmanland (Bushman) Surface 74
Busserstein 94

calcrete 49, 65, 318
caldeiro 211
canales 193
cannelure 193, 224
caprock 65, 120, 223
carbonation 54
case-hardening
case hardening – composition 250–1
case hardening – origin 250–1
case hardening and bacteria 250–1
Cassola 211
castelleted form 155, 157, *158*, 162, 165
castelleted form on dome 155, *158*
castelleted granite inselbergs 156
castle koppje see castle koppie
castle koppie 9, 113, 155, 156, *160*, *167*, 313
castle koppie – dacitic 302–3
castle koppie – definition 113, 155
castle koppie – development 156, *159*
castle koppie – distribution 113, 313
castle koppie – evolution 155–6
castle koppie and nubbin – comparison 113
castle koppie and palaeosurface *143*
castle koppie and wet site 153, 159, 190, 313
cauldron 211

cave 70, 235
cave – corestones 235–6
cave – definition 235
cave – grus 235–6
cave – plan 236, 237
cave – sheet structure *239*
chaos *269*, 275–8
chapitaux 266
chemical alteration
chemical alteration – carbonation 54
chemical alteration – clay type 54
chemical alteration – growan 54
chemical alteration – grus 54
chemical alteration – hydration 54
chemical alteration – hydrogen ion 54
chemical alteration – hydrolisis 54
chemical alteration – illite 54
chemical alteration – kaolinite 54
chemical alteration – oxidation 54
chemical alteration – plant root 54
chemical alteration – reduction 54
chemical alteration – siliceous speleothems 54
chemical alteration – solution of quartz 54
chemical alteration – solution 54
chemical alteration – water as solvent 54
Chinaman's hat: see coolie hat 186
cistern 211
classification – granite 9, 11
cleft 17, 175, *177*, *180*, 223
cliff foot cave 70, 238, *294*, *297*, *298*, *302*
climate 60
climatic azonality 309–12
climatic change 137, 293, 311, 313
climatic geomorphology 8
climatic zonality 309–12
clint and grike 316, *318*
clitter 52, 167
coarse grained 60, 162, 242
coastal forms context *298*, 313–23
coído 103
compaction 66, 181, 193, 329
compayrés 88, *89*
composition – granite 10–12, 13, 58
compressive strength 14, 101
compression 16, 34, 35, 39, 42, 43, 140, 280
concordant 10

conical font 321
conical form 159, 161, 186, 310
conical residual *71*, *161*, *162*, 165, 186
conical, or fastigiate, block 186
conjugate joints 16
conjugate shears 16
controls of weathering 58
controls of weathering – aplite *61*
controls of weathering – climate 60
controls of weathering – composition 58–62
controls of weathering – fracture density 58–62
controls of weathering – microfracture 60
controls of weathering – pegmatite 60
controls of weathering – rate of weathering 60
controls of weathering – rock texture 60
controls of weathering – texture contrast *59*
controls of weathering – time 58–9
convergence *296*
coolie hat *71*, 186
core-boulder 83, *85*, 167
corestone *51*, 58, *68,* 71, 72, 83, 84, *85*, *86*, 88, 90, 98, 167, 235, 314
corestone – basalt 99
corestone – sandstone 222, 249, 287, 296–8
corestone *in situ* 71, *88*
corestone released *316*
cottage loaf *94*
course of weathering 13, 54–8
cross joints *13*
cross-folding 122, 126
crystal cleavage 35, 56, 58
crystallisation of salts 52
Cupolakarst 298, *303*
cylindrical hollow: see rock basin

Davisian model 64
decantation gutter 224, *312*
decantation runnel *198*, 224
decompression 34
Dellen 211
demi-orange 113
dendritic drainage pattern 18
dépressions de piedemont 190
dew hole 245
diapir 10, 42
dike 10, *13*

Subject Index   345

dimpled surface 69, 71, 190, 207, 215
diorite 11
dirt scarp *1*
discordant 10
dislocated slab 41, 266–82
dislodged block 219, 264–6
displaced block/slab
displaced block/slab – origin 278
displaced block/slab – origin – compression 278
displaced block/slab – origin – fire 279
displaced block/slab – origin – insolation 248, 261, 278
displaced block/slab – origin – mineral impregnation 249
displaced block/slab – origin – quarry blasting 280
displaced block/slab – origin – seismic shaking 263, 266, 282
displaced block/slab – origin – slippage 275–8
displaced block/slab – origin – tree root growth 278
divaricating stream 71
diversion 20, 189, 224, 320
dome in gneissic granite *112*
domed inselbergs 42
domed variety 113
domes and bornhardts *118*
domes as stocks of granite intruded in gneiss 126
domes
domes – gneiss 10, 42, *45*, *145*, 255
domes – granite 132, *134*, *135*, 231
domes – structural 10, 31, 36, 126
domical granitic residuals *116*, 146
domical hill 103, 115, 126, *127*
dos de balein: see whaleback 113
dos d'elephant: see elephant rock/back 113
doughnut *229*, 230–3, 305, 319
doughnut – sandstone 305–6, *310*
drainage and faults 23
drainage patterns 20
drainage patterns – angular 18
drainage patterns – anomalous 20
drainage patterns – dendritic 18
drainage patterns – parallel 18
drainage patterns – rectangular 18
drainage patterns – straight 18, 20
drainage patterns – subparallel 18
drainage patterns – transverse 20
Druidical ceremony 211

dum-bell form 186
duricrust 49, 65
duricrust – protection 130
dwala 113
dwarf bacteria 53
dykes: see dike

egg-box type of structure 126
elephant rock 113, 125
ensiform, or blade-like, residual *186*
episodic exposure 78, 137
equifinality 106
equigranular 11, 127, 285
erosional offloading 34, 54, 98, 280
estrías 193
etch form 82, 143, 159, 168, 190
etch plain 72–4
etch surface 72, *73*, 75, 142, 201
euhedral 11
evacuation in solution 72, 149
evacuation of debris 219
evacuation of debris – anthropogenic 219
evacuation of grus 102–3
evorsion 211
exfoliation 27, 49
exhumed plain 78–9

faceted slope *170*
fan joint 16, 103
fastigiate, or conical, block 186
fault 15, 16, *21*
fault – definition – zone 96, 122
fault dislocation *3*, *272*, 281
fault scarp *1*, 77, 96, *259*, 266
fault step *41*, *42*
faulting 35, 77, 121, *123*
fault-line scarp 1, *2*, 96, 122
fault-line valley *2*, 96
Felsschüssel 211
Fernlinge 120, 140, 158
fine grained 11, 60, 125, 161, 242, 285
fire 49, *50*, 279, 285
fitted boulder 314, *315*, 322
ferricrete 49
flaggy granite 52, 218
flaggy joint 27, 218
flake 29, *50*, 58, 98, 101, 102, 238, *243*
flaking 33, 49, 50, *51*, 58, 98, 246
flared boulder *169*, 176, 179
flared slope 70, 136, 155, *157*, 173, 181, 316
flared slope – characteristics 173

flared slope – description 173
flared slope – origin 177
flared slope – overhanging 173, *174*, 184
flared slope – rhyolitic tuff *296*, 305, 309
flared slope and soil-rock junction *175*
flare slope in cleft *177*
flare slope in fracture controlled embayment 174
flare slope in subsurface 83, 177
flare slope on boulder *169*, *176*, *179*
flares after exposure 183
flask: see rock basin
flat-lying joints 13, 27
floaters 42, 83, 125
flow banding 16
flow structure *13*, 32
flowstone forms 252
fluted boulder 195
fluting 193, 194, 200, 201, 238
fluting – description 193–5
fluting – in tafone 193, 238
fluting – origin 196–200
fluting – inverted *204*
folia 12
foliation 10, 12, 42, *82*, 94
font 232, 320
font – sandstone 232–3, *320*
forest fire 285
fracture *13*, 15, 16, 27
fracture – definition 15
fracture – pattern – nonsystematic 16
fracture – pattern – random 16
fracture – pattern – systematic 16
fracture – pattern 15, 18–22
fracture – set 15, 16
fracture system 3, *15*, 16–8, 83
fracture and drainage pattern 18–23
fracture control 22
fracture density 58, 59, 78,127
freeze-thaw 52, 119, 158, 246, 263
fretted basal slope, basal fretting 184–7, *184*, 185, 186, 190, 204, 302
fretting 184, 185, 186, 190, 204, 302
fringing pediment 72, 79
frost action 70, 103, 128, 158, 313
frost riving *53*, 264
frost shattering 52, 157
fule 190

346 Subject Index

gangrenous process 58
geos 313
"ghost" 102
glacial stripping 73
glaciated granite landscape 3–4, *4*
glaciated surface 167
gnammas 211, 246
gneiss 12, *112*
gneiss dome 10, 42, *45*, *112*, *145*, 255
graben 96, 259, *260*, 266
gour 252
grain 10, 11, 14, 15, 20
granite 10
granite – characteristic forms 3, 8, 82, 294
granite – characteristics 1, 5, 16
granite – classification *9*, 11
granite – composition 10, 11, 58
granite – definition 10
granite – magmatic 9
granite – metamorphic 9, 22, 42
granite – occurrence 9
granite – origin 9, 32, 118
granite – physical characteristics 13
granite landforms – previous work 5
granite landforms – rate of development 293
granite landscape – glaciated 3, 4
granite plains 63, 81
granitoid 11
Granitkarren 173
Granitrille 224, 235, 311
granodiorite 11, *51*, 58, 73, *82*, *212*, *237*
granular disintegration 49, 98, 120, 246, 248, 249, 250
granulo 10
grike 316, *318*
groove 173, 196, 197, *199*, 213, 219, 297, 316
grotte 238
groundwater 49, 63, 66, 90, 91, 97, *102*, 121, 128, 140
growan 54
grus 54, *57*, 58, 69, 83, 84, 85, *86*, 88, *90*, 102, 103, *104*, 105, 106, 129, 132, *147*, 155, 180, 183, 201, *212*, 214, 235, 236, 245
gruss: see grus
gutter 224, 225, 226
gutter – description 224
gutter – origin 226
gutter – subsurface 227, 232
gutter – terminology 224

gutter – terminology – acanaladuras 224
gutter – terminology – cannelure 224
gutter – terminology – cleft 224
gutter – terminology – Granitrille 224
gutter – terminology – Karren 224
gutter – terminology – Kluftkarren 224
gutter – terminology – lapiaz 224
gutter – terminology – lapiés 224
gutter – terminology – Rille 224
gutter – terminology – Silikatrille 224
gutter – terminology – slot 224
gypcrete 49
gypsum 52, 248, 254, 256

half-doughnut *231*
half-orange 113
halite 52, 246
haloclasty 53, 246
hardness 13, 127
hardway 15
Härtlinge 121, 149
hieroglyph 282, *283*
honeycomb weathering 240, *241*, 248, *311*
hood 238
hoodoo rock 222
horizontal shortening and over-thrusting as the origin for wedges 125
horizontal stresses 47
hórreo 217
horst 122, 211, 259, 266, 282
hourglass-shape 186
hydration 54, 58, 87, 102, 211, 246
hydration shattering 102
hydraulic pressure 227
hydrolysis 54, 58, 87, 102, 211, 311
hydrothermal metamorphism 101
hyphae (lichen) 53, 217
hypidiomorphic 11

inheritance 20, 153
inherited drainage 23
inselberg 109, 174
inselberg – characteristics 110, 173
inselberg – definition 109, 110
inselberg – plan 113, 116, 189
inselberg – quartzitic 298
Inselberg landscape 4, *5*, 63, 79

Inselberglandschaften 4, 63, 110
insolation 33, 49, 98, 101, 130, 204, 248, 285, 287
intersection of fractures 212
intradosal zone 99
inversion 35, *38*, 183, 203
inverted Rillen *204*
iron concentration 98
island mount 109

jabre 83
joint *13*, 15, 16
joint – definition 15
joint – origin 101

kamenitza 306
Kannelierungen 193
kaolinite 54
kaolinitic speleothems 88
Karren 202, 224
kernel 83
Kluftkarren *194*, 196, 224, 227, 293, 318
knick point: see nick point
knoll 113
kociolki 211
kopje: see castle koppie
koppie: see castle koppie

laccolith 10
Lägerklufte 27
lamination 58, *85*, 98, 282
lapiaz 224
lapiés 224
lateral corrasion 71
laterite 3, 49, 74
leucocratic 11
lichens – role of 197, 217
Liesegang ring 102
lift 15
lineament and regional structure 16, 293
linear depression 190, 193, 198
lineation 12, 102, 313
lithological azonality 293
lithological contrast 77, 126, 193, 315
lithological convergence *296*
lithological zonality 293–309
loganstone 93
logging stone 93, *94*, 222
longitudinal joints *13*
lopolith 10
lowering of surface 64, 65, 66, 70, 72, 98, 233

mamillation 238, 242, *242*, 248
mantled pediment *65*, 68, 69, 71

marginal weathering 97–8, 159, 171, 298, 313
marmita 211
mass permeability 14
massif *8*, 46, 58, *70*, 71, 83, *87*, *88*, 113, 122, 139, *140*, 145, 163, 149
matopos 96, 113, 127
meda 113, 159
meda – development 161
meda and frost action 70, 103, 128, 162, 313
megaplane of subhorizontal shearing *124*
meias laranjas 113
melanocratic 11, 211
menhir 166, 185, 202, *203*, 213
meringue surface 215
mesa *4*, 72, 74, 126, *148*, 189
microfissure 15, 56, 58, 60, 63, 101
midget bacteria: see dwarf bacteria
migmatite 9, 32, 74
mineral banding 63, 81, 97, 99, 102
mineral impregnation 249
minor scarp *260*
moas 113
moat 70, 180, 190, 221, *222*
mogote 285, 286, *303*
moletes 94
monadnock 65, 120, 121, 149
monadnock de position 120
monkstones 94
monzonite 11, 29, 31, 58, 167
morros *5*, 42, 113, 125, *129*
multicyclic 75
multicyclic forms 75
multicyclic landscape 75, 116, *118*, 140–2
multiflared slope 132, 174
multi-stage development 93
multi-stage features 128
mushroom rock 186, 222, 223, *226*, 243

nannobacteria 53
navas 113
nick point 76
nigth-well 211
notch (flare) 327, *328*
nubbin 9, 113, 155, 313, 317
nubbin – definition 113, 153, 155
nubbin – development 153, *155*
nubbin – distribution 153
nubbin – evolution 153
nubbins and castle koppie – comparison 15, 159

nuclear explosion 49
nunatak 159

obduction 9
offloading 27, 33, 34, 35, 39, 46, 54, 98, 278, 280
onion-skin weathering 98
opal A dissication cracks *253*
opal A 253, *253*, *254*
opaline speleothems 88
Opferkessel 211, 219
ordered water 102
oriçangas 211
orogen 9, *10*, 22
orthogonal fracture *3*, *14*, 16–8, 29, *30*, 42, 48, 63, 83, 154, 155
orthogonal system 16, 28, 32, 35, 63, 116, 278, 287
overhanging flared slope 173, *174*
overhanging wall 173, 184, 193, *196*, 197, 216
overlapping slabs 271, 274, 280
overthrusting *36*, *40, 43*, 125
oxidation 54

palaeosurface 140, *143*, *154*, 155, 313
pan: see rock basin
parallel drainage pattern 18
parted block 264, 266
pecking 322, 324
pedestal rock 186, 222, 306, *311*
pedestal rock – description 222
pedestal rock – origin 222–23
pedestal rock – sandstone 222
pedestal rock – terminology 222
pedestal rock – terminology – hoodoo rock 222
pedestal rock – terminology – mushroom rock 222
pedestal rock – terminology – Pilzfelsen 222
pedestal rock – terminology – rocas fungiformes 222
pedestal rock – terminology – roches champignons 222
pedestal rock – terminology – Tischfelsen 222
pediment 66, 70, 71, 72, 74
pediment – characteristics 66
pediment – definition 66
pediment – origin 66
pediment – origin – lateral corrasion 71
pediment – origin – mantle controlled planation 68, 71
pediment – types of 66, 71

pediment – types of – covered 66, 71
pediment – types – mantle 66, 68, 69, 71
pediment – types of – rock 66, 71
pediment and peneplain 64, 65, 72, *296*
pediment nick 66
pediplain 74
pediplanation 74
pegmatite 11, *60*
penas abaladoiras 93
peneplain 64, 65, 72
peneplain – lithology 64, 65, 72
penitent rocks 94, *95*
perched block *92*, 94
perched slab *278*
peripheral weathering
peripheral weathering causes – erosional offloading 98
peripheral weathering causes – insolation 98, 101
peripheral weathering causes – metamorphism 99, 101
peripheral weathering causes – pressure release 99, 101
peripheral weathering causes – primary petrology 99
peripheral weathering causes – shearing 101
peripheral weathering causes – volume increase 101
permeability 14. 15, 56, 72, 98, 231, 248, 329
perviousness 15, 60
phacolit 10, 130
phased development 155
physical characteristics 13
physical disintegration 49–54, 217
pía 211
piedmont 64, 66, 68, 69, 70, 106, 110, 120
piedmont angle 66, 69, 70, 72, 110, 120, 149, *188*, 188–90, 193
piedmont knick: see piedmont nick
piedmont nick 66, 110, *188*, 188, 193
pilancón 211
Pilzfelsen 222
pillar 128, *136*, 222, *296*
pit: see rock basin
pitted surface 54, *55*, 56, 71, 199, *212*, 227
pitting 98
pitting – limestone *297*

348 Subject Index

plain 63
plain – Adamaua 63
plain – Banda 63
plain – epigene 63
plain – granite 63, *64*
plain – Kordofan 63
plain – peneplain 64, 72
plain – pediment 63, 72
plain – rolling 64, 72
plain – Rovuma 63
plain – subaerial 63
plain – very flat 74, *75*
planation surface 64, 65, 72, 74, 77, *118*, *119*, *296*
plate tectonics 9
platform 69, 70, 71, 98, 190
platform – structural 74
plinth 92, 220
plinth – description 220–21
plinth – origin 221
pluton 9, 10
poço 211
polygonal cracking 282, 283, 284, 285, 286
polygonal cracking – description 282–4
polygonal cracking – origin 285, 286, 287, 288, 289, *290*
polygonal cracking – origin –increase in volume 280, 289
polygonal cracking – origin – mineral concentration 287, 289
polygonal cracking – origin –shearing 287
polygonal cracking – origin – weathering front 285, 287, 289
polygonal cracking – previous interpretations 285–7
polygonal cracking – previous interpretations – dessication 288
polygonal cracking – previous interpretations – extension 287
polygonal cracking – previous interpretations – insolation 285, 287, 289
polygonal cracking – previous interpretations – shearing 287
polygonal cracking on corestones *286*, 289
polygonal cracks 283, *286*, 287, 289
pop-ups: see A-tents
porosity 14
porphyry 11, *12*, 127, 184, 197, 285
porphyritic granite 12
positive feedback 58, 129, 145, 329

pot-hole 199, 211, 225, 320
primary petrological structures 99
primary permeability 14
pressure release 27, 33, 34, 45, 99, 280
pseudobedding 27, 28, 52, *94*, *166*, 313
pseudokarren 193, 197, 224
pseudokarst 235
pseudostratification 27

quarry blasting 280
quartz in solution 54
quartz veins 81
quartzitic dome 298

radial drainage patterns 19
radiating joint 16
rapids 76
rectangular drainage pattern 18, 19
reduction 54
regolith 47, 49, 56, 69, 119, 149, 181, 183
regueros 193
reinforcement 22, 58, 129, 145, 329
rejuvenation head 75, 135
remnants of circumdenudation 120, 121, 232
rhomboidal pattern 16
rhomboidal system 16
rib 60, 82, 110, 183, 203, 242, 250
Riefelungen 193
rift 13, 15, 32, 262
Rillen 168, 219, 224
Rillen, inverted 204
rim 3, 197, 211, 224, 232
ring complex 10
rocas fungiformes 222
roches champignons 222
rock basin 167, 207, 309, 320
rock basin – description 207
rock basin – development 211, 219, 232, 321
rock basin – differentiation 211, 216
rock basin – irregular shape 207, 212, 214
rock basin – nomenclature 211
rock basin – nomenclature – araceenhorst 211
rock basin – nomenclature – bath tub 211, 214
rock basin – nomenclature – Baumverfallspingen 211

rock basin – nomenclature – caldeiro 211
rock basin – nomenclature – cassola 211
rock basin – nomenclature – cauldron 211
rock basin – nomenclature – cistern 211
rock basin – nomenclature – Dellen 211
rock basin – nomenclature – Felsschüssel 211
rock basin – nomenclature – gnamma 211
rock basin – nomenclature – granite pit 211
rock basin – nomenclature – kociolki 211
rock basin – nomenclature – marmita 211
rock basin – nomenclature – nigth-well 211
rock basin – nomenclature – Opferkessel 211, 219
rock basin – nomenclature – oriçanga 211
rock basin – nomenclature – pía *211*
rock basin – nomenclature – pilancón 211
rock basin – nomenclature – poço 211
rock basin – nomenclature – rock bason 211
rock basin – nomenclature – rock hole 211
rock basin – nomenclature – tanque 211
rock basin – nomenclature – vasque rocheuse 211
rock basin – nomenclature – Verwitterungsnäpfe 211
rock basin – nomenclature – water eye 211
rock basin – nomenclature – weather pit 211
rock basin – origin 211
rock basin – origin – anthropogenic 211, 219
rock basin – origin – deflation 219
rock basin – origin – epigene 213
rock basin – origin – flushing 219
rock basin – origin – gravity 211, 213
rock basin – origin – solution 211, 219

Subject Index 349

rock basin – origin – structural 212, 216
rock basin – origin – subsurface 213, 223
rock basin – origin – weathering 211
rock basin – sandstone 222, 240, 249
rock basin – size 207, 219
rock basin – topographic distribution 210, 218
rock basin – types 207, 216, 218
rock basin – types – armchairshaped- hollows 207, *209*, 218
rock basin – types – cylindrical hollow 207, *209*, 219
rock basin – types – flask 207, 216, 219, 227
rock basin – types – pan 207, *208*, *212*, *214*, 216, 218
rock basin – types – pit 207, 208, 216, 218
rock bason 211
rock doughnut *229*, 230
rock doughnuts – description *229*, 230
rock doughnuts – origin 229, 230, 231
rock hole 211
rock levee 228, 229
Rock levee – origin 228–32
rock pediment 66, 69, 72, 190, 224, 308
rock platform 69, 190
rock platform – description 69, 190
rock platform – origin 69, 190
rock texture 60
rock texture – allotriomorphic 11
rock texture – coarse grained 60, 162, 242
rock texture – equigranular 11, 127, 285
rock texture – fine grained 11, 60, 125, 161, 242
rock texture – hypidiomorphic 11
rock texture – porphyritic 11, 127, 184, 197
rock texture – subidiomorphic 11
rock texture – xenomorphic 11
role of lichens 217, 218
rolling plain 49, 64, *64*, 75
rounding 83, 86, 120, 329
Rundkarren 193
runnel: see gutter
ruware 113

saibro 83
salt crystallisation 52, 246, 248
Salzsprengung 248
saucer-shapped depression 190, *215*, 216
sauló 83
scales 16, 183, 242, 329
scalloped 240, 242
scarp foot and flare 70, 135, 136, 142, *176*, 190
scarp retreat 65, 69, 74, 120, 130, 137, 142
scarp retreat and pedimentation 65, 69, 74, 120, 140
scarp-foot depression 70, 190
scarp-foot depression – description 70, 190
scarp-foot depression – origin 70, 190, *192*
scarp-foot depression – origin – structural 192, 193
scarp-foot depression – origin – volume decrease 193
scarp-foot depression – origin – weathering in scarp-foot 193
scarp-foot erosion 188, 193, 239
scarp-foot weathering 188, 193
scree 52, 159, 166, 329
sea cave 313
set 15
schistosity 12
secondary permeability 15
seismicity 263, 282
shear 16, 39, 140, 282
shear plane 16, 97
shear stress 16, 288
shearing 17, 43, 101, *124*, 126, *130*, 140, 287
sheet fractures: see also sheet structure 16, 30, 31, 32, 33, 43, 63, 104, 154
sheet structure 28, 30, 313
sheet structure – characteristics 28
sheet structure – characteristics – age 35, 39
sheet structure – characteristics – climatic distribution 32
sheet structure – characteristics – depth 27, 29, 30
sheet structure – characteristics – lithology 33, 72
sheet structure – characteristics – parallel to surface 30, 34, 41
sheet structure – characteristics – radious of curvature 30
sheet structure – characteristics – relation to other structure 32

sheet structure – characteristics – thickness 29
sheet structure – characteristics – variation in depth 29, 30, 33
sheet structure – definition 28
sheet structure – description 28
sheet structure – en echelon pattern 29
sheet structure – endogenetic explanations 41
sheet structure – endogenetic explanations – concentric structure 42
sheet structure – endogenetic explanations – diapir 42
sheet structure – endogenetic explanations – lateral compression 43
sheet structure – endogenetic explanations – magmatic stress 41
sheet structure – inverted relief 27, 35, 38
sheet structure – synform in dome 35, *36*, *37*, 38
sheet structure – terminology 27
sheet structure – terminology – Bankung 27
sheet structure – terminology – estructura en capas 27
sheet structure – terminology – exfoliation 27
sheet structure – terminology – flat-lying joint 27
sheet structure – terminology – Lagerklufte 27
sheet structure – terminology – offloading joint 27, 33
sheet structure – terminology – pressure release joint 27, 33, 45
sheet structure – terminology – relief of load joint 27
sheet structure – terminology – stretching plane 27
sheet structure – terminology – structure en gros bancs 27
sheet structure – theories of origin 32
sheet structure – theories of origin – consequent on surface form 32
sheet structure – theories of origin – exogenetic 33
sheet structure – theories of origin – exogenetic – chemical weathering 33

350 Subject Index

sheet structure – theories of origin – exogenetic – insolation 33
sheet structure – theories of origin – exogenetic – offloading – difficulties 33, 34, 35
sheet structure – theories of origin – exogenetic – offloading 33, 34, 35
sheet structure – theories of origin – exogenetic – pressure release 33, 45
sheet structure – theories of origin – primary feature of the rock 33
sheeting – youthful 36
shelter 238, 249
shield 8, 10, 42, 63, 75, 149, 176, 183
shield lands 8, 62, 63, 72, 149
shingle beach 103, 315
shore platform 315, 322
sial 9
silcrete 3, 49, 74, 145
silica solution 87, 88
silicate karst 235
siliceous speleothems 54, 250, 251–6
Silikatkarren 193
Silikatrille 193, 224, 235
sill 10, 13, 32, 97, 264
slipped slab *275*, *276*, 280, 282
sliver *50*, 98, 278
slot 17, 167, 224, 298, *303*
solution 54, 65, 66, 72, 87
solutional evacuation 72, 149, 167
source of salts 248
spalling 33, 49, *50*, *51*, *52*, 58, *91*, 98
specific gravity 13
speleothem 251, 252
speleothem types 252
speleothem phases 253, 254
speleothem – gypsum tip 254, *255*, 256
speleothem – gypsum whisker 254, *255*, 256
speleothem – opal-A 251, 252, 253, *254*, *255*, *256*
spheroidal weathering 84, 97
split rock 259
split rock description 259
split rock – development 259–60, 264
split rock – origin 260
split rock – origin – gravity 262, *263*, 264

split rock – origin – insolation heating and cooling 260, 278, 279, 289
split rock – origin – seismicity 263, 266, 282
steeply inclined fracture 18, *32*, 103, *117*, 149, *159*, 162, 264, 313
stepped inselbergs 78, 132–8, *138*
stepped topography 77–8
stepped topography and weathering 77–8
stock 10, 97, 126, *127*
straight drainage pattern 18–23
strandflat 322, *323*
stream persistence 20
stress 15, 16, *18*, 33, 34, 35, 39, 43, 46
stretching plane *13*, 27, 43
subhedral 11
subidiomorphic 11
subparallel drainage pattern 18
subsurface 83, 228, 264
subsurface flushing 72
subsurface initiation 131–2, 200, *309*
subsurface weathering 79, 88, 90, 91, 93, *104*, 127–30
superimposition 20
swamp slot 298, *303*
syenite 11, 197
synform 35, *36*, *37*, *38*, *129*, 327
systems 15

tafone 53, 70, 237, 310, 311
tafone – boulder 237–8
tafone – definition 238
tafone – development 238, *241*, 243, 246, *249*
tafone – development – freeze-thaw 240, 246
tafone – development – haloclasty 246, 249
tafone – development – humidity 246
tafone – development – salt crystallisation 246, 248
tafone – development – load concentration 243, 246
tafone – development – load concentration – edaphical weathering *247*
tafone – development – load concentration – strain: lacunar zones 246, *247*
tafone – initiation 245–6
tafone – stages of development 249–50

tafone – temperature variation 246
tafoni 238–51, 310
tanque 211
tectonic and structural forms differentiated 95–7, 121
tectonic region *10*
tensile strength 14, 101
tension fracture 16, *275*
tesseleted pavement *290*, 291
tethraedral cornerstones 101
time 62
Tischfelsen 222
tombstones 94
tonalite 11
tor 42, 147, 155
tortoiseshell rock 243
tower 162–3, *296*, *312*
towerkarst 298
transverse drainage pattern 20, *22*
travertine *67*
tree root, disruption by *53*, 278, 280
triangular divide 183
triangular lobe 183
tubular 235, *238*
Turmkarst 298
turtleback 113, 130
two stage concept 47, 81, *88*, 91, 127–30, 168
two stage development 81, *89*, 91, 105, 106, 149

unbuttressing 264
unconformity 78
undulating 64

valley impression 20
valley in valley forms 76
valley side facets 76
Variscan orogeny 22
vasque rocheuse 211
vein 2, 60, *61*, 81, 84, *87*, 93, 97, 197, 236
vertical wedge 125, 267, *269*, *270*, 278
Verwitterungsnäpfe 211
visor 238, *240*, 245, 249
volcanic plug 298
volume decrease 66
volume increase 33, 58, 101, 102, 280

water eye 211
water table in scarp foot 161, 176, 178
waterfall *23*, 76
weak karst 49

Subject Index 351

weather pit 211
weathering 49, 54, 56, 63
weathering – bauxite 49
weathering – calcrete 49
weathering – cementation 49
weathering – definition 49
weathering – duricrust 49
weathering – ferricrete 49
weathering – from surface downward 56
weathering – gypcrete 49
weathering – induration 49
weathering – karst 49
weathering – laterite 49, 74
weathering – significance 49
weathering – silcrete 49
weathering along sheet structure 54, *104*
weathering and plains 63–4
weathering and stepped topography 77
weathering front 49, 56, 71, 72, 136, 183
weathering front – mineral accumulation 49
wedge *30*, *32*, *39*, *40*, 276
wedge – description 39, 40, 276–82
well-rounded boulder *84*
wetting and drying 161, 204

whaleback *112*, 113
whiskers druse tip 254
whiskers of gypsum in speleothems 254

xábrego 83
xenoliths 211, 249
xenomorphic 11

yelmos 113

Zeitschrift für Geomorphologie – special inselberg issue 5–8
zlobki 193
zonality 293, 309–12

About the authors

Charles Rowland Twidale (1930) obtained his Doctoral degree in Geology from the University of Bristol (1957) and is Honoris Causa at the Complutense University of Madrid (1983). At present he is Emeritus Professor at the University of Adelaide, South Australia, and Research Leader of the University Institute of Geology "Isidro Parga Pondal" of the University of Coruña. He has worked on geomorphology subjects in North America, Australia, Africa and Europe to gain insight in structural geomorphology, granitic geomorphology, eolian deposits in deserts areas and etching processes in continental and marine environments. In addition, he worked on the development of models of landscape evolution in intraplate continental areas. As a specialist in granitic geomorphology, he has published numerous papers in international scientific journals and numerous books on this topic and on landscape evolution. Twidale is the author of *Granite Landforms* (1982, Elsevier, Amsterdam).

Juan Ramón Vidal Romaní (1946) obtained his Doctoral degree in Geology from the Complutense-University of Madrid (1983). He is a Professor in Geodynamics at the University of Coruña and Director of the University Institute of Geology "Isidro Parga Pondal". He has worked on granitic geomorphology and on its relation to the particular characteristics of landscapes, like glacial, coastal and continental landscapes, either in past or present climates. He has developed new research methods for cosmogenic chronology, erosive granite surfaces, granitic pseudokarst processes and for the genesis of the granitic forms. By field work in Argentina, Australia, Madagascar, Portugal, Spain and North African countries, he has become a specialist in the interpretation of the origin of granitic forms in relation to their geodynamic environment.